Taguchi Techniques for Quality Engineering

Taguchi Techniques for Quality Engineering

Loss Function, Orthogonal Experiments,
Parameter and Tolerance Design

Phillip J. Ross

McGraw-Hill Book Company

New York St. Louis San Francisco Auckland
Bogotá Hamburg London Madrid Mexico
Milan Montreal New Delhi Panama
Paris São Paulo Singapore
Sydney Tokyo Toronto

Library of Congress Cataloging-in-Publication Data

Ross, Phillip J.
  Taguchi techniques for quality engineering/Phillip J. Ross.
    p. cm.
  ISBN 0-07-053866-2
  1. Quality control—Statistical methods. 2. Engineering design—
Statistical methods. I. Title.
TS156.R67 1988
620′.0045—dc 19   88-1701
                                                                      CIP

    567890 DOC/DOC   93210

ISBN 0-07-053866-2

*The editors for this book were Betty Sun and Dennis Gleason, the
designer was Naomi Auerbach, and the production supervisor was
Suzanne Babeuf. It was set in Century Schoolbook by Professional
Composition, Inc.*

Printed and bound by R. R. Donnelly & Sons Company.

To Emily,
who is always first in my book

A chasm.
A bridge to build.
Not as a monument in itself,
but to make it easier for others to follow.

Contents

Preface   xi

Introduction   xiii

Chapter 1. The Economics of Reducing Variation   1

1-1   The Meaning of Quality   1
1-2   Goalpost Philosophy   2
1-3   Taguchi Loss Function   3
1-4   Comparison of Philosophies   3
1-5   Case Study: Polyethylene Film   5
1-6   Japan's Desire for Low Loss   7
1-7   Case Study: Tool Wear in a Process   7
1-8   Case Study: Automatic Transmissions   16
1-9   Factory Tolerance   17
1-10   Other Loss Functions   17
1-11   General Loss Function for Nominal-Is-Best Situation   18
1-12   Summary   20

Chapter 2. Introduction to Analysis of Variance   23

2-1   Understanding Variation   23
2-2   No-Way ANOVA   24
2-3   One-Way ANOVA   31
2-4   Two-Way ANOVA   47
2-5   Three-Way ANOVA   54
2-6   Critique of the F Test   60
2-7   Summary   61

Chapter 3. Introduction to Orthogonal Arrays   63

3-1   Typical Test Strategies   63
3-2   Better Test Strategies   67
3-3   Efficient Test Strategies   68
3-4   Steps in Designing, Conducting, and Analyzing an Experiment   71
3-5   Example Experimental Procedure: Popcorn Experiment   95
3-6   Summary   98

Chapter 4. Multiple-Level Experiments   101

4-1   Necessity for Multiple-Level Experiments   101
4-2   Conversion from Two to Four Levels   101
4-3   ANOVA for Four-Level Factors   103
4-4   Polynomial Decomposition   106
4-5   Multiple-Level Factor Interactions   109
4-6   Dummy Treatment for Three-Level Factors   109
4-7   ANOVA for a Dummy Treatment   110
4-8   Summary   112

Chapter 5. Interpretation of Experimental Results   115

5-1   Interpretation Methods   115
5-2   Percent Contribution   116
5-3   Estimating the Mean   118
5-4   Confidence Interval around the Estimated Mean   120
5-5   Omega Transformation   124
5-6   Other Interpretation Methods   125
5-7   Summary   131

Chapter 6. Special Designs   133

6-1   Nested Experiments   133
6-2   Combined Factors   137
6-3   Idle Column Method   142
6-4   Summary of Multiple Levels   145
6-5   Component Search   146
6-6   Summary   149

Chapter 7. Attribute Data   151

7-1   Introduction to Attribute Data   151
7-2   Sample Size for Attribute Data   153
7-3   Analysis of Attribute Data   153
7-4   Case Study: Casting Cracks and Two Classes   154
7-5   Casting Cracks with Multiple Classes   161
7-6   Two-Class Attribute Data; Known Occurrences Only   164
7-7   Summary   165

Chapter 8. Parameter and Tolerance Design   167

8-1   Parameter and Tolerance Design Explanation   167
8-2   Control and Noise Factors   168
8-3   Introduction to Parameter Design   170
8-4   Signal-to-Noise Ratios   172
8-5   Parameter Design Strategy   175
8-6   Case Studies of Parameter Design   176
8-7   Analysis of Inner/Outer Array Experiment   184
8-8   Alternative Inner/Outer OA Experiment   191
8-9   Measurement System Parameter Design   195
8-10   Tolerance Design   201
8-11   Quality Countermeasures   202

8-12   Steps in Experimentation   203
8-13   Summary   205

Appendix A. Problem Solutions   **207**

Appendix B. Two-Level Orthogonal Arrays   **213**

Appendix C. Three-Level Orthogonal Arrays   **227**

Appendix D. Design and Analysis Tables   **233**

Appendix E. Proofs   **247**

Appendix F. Definitions and Symbols   **251**

Appendix G. Example Experiments   **257**

Index   277

Preface

Taguchi Techniques for Quality Engineering is intended as a guide and reference source for industrial practitioners (managers, engineers, and scientists) involved in product or process experimentation and development. Most engineers are familiar with setting up tests to model actual field conditions and the cause-effect relationship of design to performance; however, their knowledge of a proper testing strategy is usually limited. When engineers have had exposure to experimental design, typically, the reaction is to deem the approach too costly and time consuming because of full factorial designs.

It is most unfortunate that people are not aware of the potential savings in test time and money offered by more efficient testing strategies. Not only are savings in test time and cost available but also a more fully developed product or process will emerge with the use of better experimental strategies.

The Taguchi philosophy provides two tenets: (1) the reduction in variation (improved quality) of a product or process represents a lower loss to society, and (2) the proper development strategy can intentionally reduce variation. Again, most managers and engineers are not aware of the economics of improved quality and the techniques to achieve higher quality at lower costs.

This book addresses the basic testing and development strategies that have allowed some Japanese companies to successfully become world economic competitors. Many Japanese engineers since the mid-1960s have had Taguchi training. In 1985 Nippon Denso, a Toyota affiliate, ran over 2500 experiments concerning automotive electrical products, for example.

By using and understanding the Taguchi methods, managers and engineers will realize what is required to put western product development and quality back into the competition. Today's managers and engineers must have a certain amount of exposure to these methods before they can appreciate how much improvement in testing and development strategies can be made. The text takes a user-oriented, hands-on approach for working engineers or scientists and their immediate management to develop initial expertise in the Taguchi methodology. There are currently few sources of information about Taguchi methods and these speak, typically, from a statistician's point of view. This book should be quite useful to the statistically inexperienced engineer who would have some difficulty understanding and utilizing a traditional text concerning designed experiments.

It is hoped that *Taguchi Techniques for Quality Engineering* will bridge the gap for the industrial user, eventually making American companies more competitive with their products in a world market. World-class quality will be a requirement for corporations to remain lucrative, let alone highly competitive, as more and more companies embrace the Taguchi methods.

My indebtedness is to Bill Diamond who introduced me to experimentation based on orthogonal arrays, the Hadamard matrices. I know he would hate to admit it, but the Taguchi matrices are mathematically and statistically equivalent to Hadamard's. The approaches used by the two experimenters may be different but the matrices are not.

I would like to give my thanks to Kathy Layne and Steve Abney, two very dear friends, who acted as statistical consultants, an editorial staff, and sympathetic ears during the creation of this text. My thanks also to those who suffered through some of the pilot classes and seminars on designed experiments; your contribution has been the engineer's point of view and, for me, a better appreciation of what the customer expects from this product. And lastly, thanks to my family for the tolerance of all the lost evenings and weekends as Dad lingered over the PC.

The author expresses gratitude to the American Supplier Institute, Inc., Center for Taguchi Methods, for granting permission to reproduce the orthogonal array, triangular table, and related linear graphs. They are contained in ASI's new (1987) edition of *Orthogonal Arrays and Linear Graphs*.

Phillip J. Ross

Introduction

Taguchi addresses quality in two main areas: off-line and on-line quality control (QC). Both of these areas are very cost sensitive in the decisions that are made with respect to the activities in each. Off-line QC refers to the improvement of quality in the product and process development stages. On-line QC refers to the monitoring of current manufacturing processes to verify the quality levels produced. The off-line portion of QC is addressed in this text because of the paucity of materials on this phase of Taguchi methods and the positive impact on cost that is obtained by improving quality at the earliest times in a product life cycle.

The on-line phase is covered by many texts as a dimensional approach to quality control, typified by statistical process control, or SPC, and for this reason will not be addressed in this text. The Taguchi on-line QC approach is a cost quality control perspective and some day should be recognized as an alternative quality control system.

This text reviews the basic aspects of off-line QC developed by Taguchi. There are several more sophisticated concepts for off-line QC that are not covered in this text but should be pursued by the experimenter after initial work with these methods. Also, education and training in general statistical methods is recommended. In particular other designed experiment texts will be valuable for the experimenter as background. A short description of the chapter contents follows.

Chapter 1. The Economics of Reducing Variation

The economics (cost reduction) of reducing variation is a subject that is not addressed widely at this time. Taguchi uses a different cost model for product characteristics than is typically used, which places more emphasis on reducing variation, particularly when the total product variation is within the specification limits for the product. This chapter covers the conventional viewpoint of cost versus specification limits and introduces the Taguchi model for cost versus specification limits, the loss function. The Taguchi methodology ascribes to the approach that the lowest loss to society represents the product with the highest quality. Higher product quality by definition means less variation of a product characteristic. The difference between conventional and Taguchi approaches to higher quality is that one proposes that higher quality costs more and the other proposes that higher quality costs less. The loss function is a mathematical way of quantifying the cost as a function of product variation which answers the question of whether further reduction of variation will reduce costs.

Also discussed is the difference between off-line and on-line QC and where the responsibility falls for that portion of total quality control. The different types of loss functions are described for the typical types of product characteristics such as higher is better, nominal is best, and lower is better.

Chapter 2.   Introduction to Analysis of Variance

Chapter 2 introduces the basics of conducting an analysis of variance for a particular set of data and how the ANOVA techniques work from a statistical basis. The examples are extremely simple and easy to follow because the purpose of the book is to teach a graduate person the mechanics of ANOVA, not to practice basic arithmetic. The F test, a basic statistical test of comparisons of products, is introduced and explained in a simple manner as a tool to make decisions after an ANOVA is completed.

Chapter 3.   Introduction to Orthogonal Arrays

Orthogonal arrays are introduced from the viewpoint of the pragmatist who is always trying to make product improvement decisions with the minimum amount of test data. Using minimal amount of test data is not necessarily a problem in itself; however, the considerations of what may make up a valid experiment from a risk viewpoint are seldom considered by the typical experimenter. The statistical aspects of the size of an experiment are discussed in conjunction with the amount of information required to be evaluated during the experiment. The aspects of designing a simple orthogonal experiment are discussed, including the handling of various factors and interaction effects. Also discussed are practical randomization techniques and how they apply to orthogonal array experimentation; a technique called *blocking*, which allows known sources of variation to be controlled in an experiment rather than depending upon randomization; and how to make an estimate of experimental error.

Chapter 4. Multiple-Level Experiments

Some special methods of modifying a two-level orthogonal array to accommodate multiple (more than two) level factors are introduced and limitations of each are discussed. Polynomial decomposition of variance is introduced with a discussion of what information is gained when running a multiple-level versus a two-level experiment.

Chapter 5. Interpretation of Experimental Results

Methods of estimating different values such as the percent contribution, the mean, and confidence intervals from the experimental data are discussed. Two methods of estimating the mean are used, standard and omega transformation. Other interpretation methods of lesser statistical sophistication are discussed also.

Chapter 6. Special Designs

A typical situation that confronts an engineer is handled with nested experiments. Many times discrete variables such as different materials cause a portion of the experiment to be unable to be exposed to the variation of other factors. This situation requires the use of a nested experiment to allow easy analysis of the data. Engineers will find themselves running into this fairly frequently. Two methods of handling a mixture of factors with different numbers of levels are covered and a component search technique is described for problems where retest is possible. Many nondestructive tests could benefit from this test strategy.

Chapter 7. Attribute Data

All of the problems discussed up to this point concerned variable data, which are on a continuous scale from high to low, such as weight, diameter, and voltage. However, some experiments do not lend themselves to this type of measurement system. Results are on a discontinuous scale such as good or bad, passing or failing, meets specification or doesn't meet specification. Such ratings become attribute characteristics and require a different type of analysis than variable data. The factors may be allocated to an experiment in the same manner, but the analysis can be modified for attribute data.

Chapter 8. Parameter and Tolerance Design

The main thrust of Taguchi methods is the use of parameter design which is the ability to design a product or process to be resistant to various environmental factors that change continuously with customer use. To determine the best design of a product or process requires the use of a strategically designed experiment which exposes the product or process to the varying environmental conditions. Taguchi refers to these variations in customer use as noise factors. The analysis of the experimental results uses a signal-to-noise ratio to aid in the determination of the best product or process designs. Nonlinear response characteristics of products or processes can be used to the engineer's advantage if the proper design philosophy is used.

Taguchi Techniques
for Quality
Engineering

The Economics of Reducing Variation

1-1 The Meaning of Quality

Products have characteristics that describe their performance relative to customer requirements or expectations. Characteristics such as fuel economy of a car, the weight of a package of breakfast cereal, the power losses of a home hot water heater, or the breaking strength of fishing line are all examples of product characteristics that are of concern to customers at one time or another.

The quality of a product is measured in terms of these characteristics. Quality is related to the loss to society caused by a product during its life cycle. A truly high quality product will have a minimal loss to society as it goes through this life cycle. The loss a customer sustains can take many forms, but it is generally a loss of product function or properties. Other losses are time, pollution, noise, etc. If a product does not perform as expected, the customer senses some loss. After a product is shipped, a decision point is reached; it is the point at which the producer can do nothing more to the product. Before shipment the producer can use expensive or inexpensive materials, use an expensive or inexpensive process, etc.; but once shipped, the commitment is made for a certain product expense during the remainder of its life.

Quality has but one true evaluator, the customer. A "quality circle" that describes this situation is shown in Figure 1-1. The customer is judge, jury, and executioner in this model. Customers vote with their wallets on which products meet their requirements, including price and performance. The birth of a product, if you will, is when a designer takes information from the customer (market) to define what the customer wants, needs, and expects from a particular product. Sometimes, a new idea (high technology) creates its own market, but once a competitor can duplicate the product, the technological advantage is lost.

The designer must take the customer's wants, needs, and expectations and translate them into product specifications, which include drawings, dimensions, tolerances, materials, processes, tooling, and gaging. The makers use this information, along with the prescribed machinery, to fabricate the product. The product is then delivered via marketing channels to the customer. To satisfy the customer the product must arrive in the right quantities, at the right time, at the right place, and provide the right functions for the right period of time. All of this must be available to the customer at the right price, too. This is a tough order to fill, but the simplest definition of high quality is a happy customer. Customers should become more endeared to a product the more it is used. Customer feedback to the designers and makers comes in terms of the number of products sold and the warranty, repair, and complaint rate. Increasing sales volume and market share with low warranty, repair, or complaint rates translates to happy customers.

1-2 Goalpost Philosophy

Today in America it is quite popular to take a very strict view of what constitutes quality. In his book, Crosby supports the position that a product made according to the print, within permitted tolerance, is of

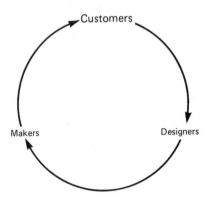

Figure 1-1 Quality circle.

high quality.* This strict viewpoint embraces only the designers and the makers. This is the "goalpost" syndrome.† What is missing from this philosophy is the customer's requirements. A product may meet print specifications, but if the print does not meet customer requirements, then true quality cannot be present. For example, customers buy TVs with the best picture, not ones that necessarily meet specifications.

Another example showing that the goalpost syndrome contradicts the customer's desires is as follows. Batteries supply a voltage to a light bulb in a flashlight. There is some nominal voltage, let us say 3 volts, that will provide the brightest light but will not burn out the bulb prematurely. Customers want the voltage to be as close to the nominal voltage as possible, but battery manufacturers may be using a wider tolerance than allowed by the battery specification. So, as a result, some flashlights burn dimly and others burn out the bulbs prematurely. Customers want the product close to nominal all the time, and producers want to allow the product to vary to the limit of the specifications; how can these seemingly incongruent ideas be brought into harmony?

1-3 Taguchi Loss Function

The Taguchi loss function recognizes the customer's desire to have products that are more consistent, part to part, and a producer's desire to make a low-cost product. The loss to society is composed of the costs incurred in the production process as well as the costs encountered during use by the customer (repair, lost business, etc.). To minimize the loss to society is the strategy that will encourage uniform products and reduce costs at the point of production and at the point of consumption. Let's look at a comparison of the goalpost and loss function philosophies with an example.

1-4 Comparison of Philosophies

When the hood of a typical automobile is opened, a mechanism may be in place which automatically holds the hood in the open position. The force required to close the hood from this position is important to the customer. If the force is too high, then a weaker individual may have difficulty in closing the hood and ask for the mechanism to be adjusted. If the force is too low, then the hood may come down when a gust of wind hits it, and again the customer will ask for it to be adjusted. The engineering specifications and detail and assembly drawings call out a

---

*Philip B. Crosby, *Quality Is Free*. McGraw-Hill, New York, 1979.

†In football, a team is awarded three points for a field goal regardless of where the ball passes through the uprights, whether exactly midway between the uprights or far to left or right. In other words, there is a wide tolerance.

particular range of force values for the hood assembly. A range must be used, since all hoods cannot be exactly the same; a lower limit (LL) and an upper limit (UL) are specified. If the force is a little high or low the customer may be somewhat dissatisfied but may not ask for an adjustment to the hood. A goalpost view of this situation is shown graphically in Fig. 1-2.

The goalpost philosophy says that as long as the closing force is within the zone shown as the customer's tolerance, this would be satisfactory—no problem. If the closing force is smaller than the lower limit or larger than the upper limit of the customer's tolerance, then the hood would have to be adjusted at some expense, say $50, to be borne by the manufacturer (warranty).

As a customer, the closer the closing force is to the nominal, or target, value, the happier you are. If the force is a little low or a little high, you sense some loss. If the force is even lower or higher, you would sense a greater loss; the hood would come down more frequently or be uncomfortably hard to close. When the force reached the customer tolerance limits, the typical customer would complain about the hood. But what is the real difference between a closing force indicated by points A and B on the goalpost graph? It appears from a producer's viewpoint that the difference is the total cost of adjustment. From a customer's viewpoint there is very little difference in a hood that falls down just a little bit more easily. A better model for the cost versus closing force is shown in Fig. 1-3.

This curve, the loss function, more nearly describes the real situation. If the closing force is near the nominal value, there is no cost or very low cost associated with the hood. The farther the force gets from the nominal force, the greater the cost associated with that force is, until the

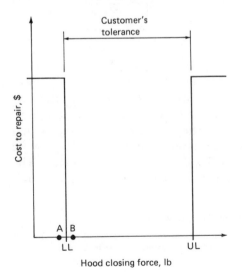

Figure 1-2 Goalpost syndrome.

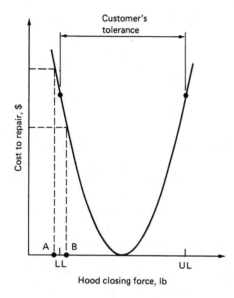

Hood closing force, lb

Figure 1-3   Taguchi loss function.

customer's limit is reached where the cost equals the adjustment cost. This model quantifies the slight difference in cost associated with a hood closing force of force *A* and force *B*. This is a fictitious example, but let's look at an actual case from Japan.

1-5   Case Study: Polyethylene Film*

A supplier in Japan made a polyethylene film with a nominal thickness of .039 in (1.0 mm) that is used for greenhouse coverings. The customers want the film to be thick enough to resist wind damage but not too thick to prevent passage of light. The producers want the film to be thinner to be able to produce more square feet of material at the same cost. A plot of these contradictory desires is shown in Fig. 1-4. At the time the national specifications for film thickness stated that the film should be .039 in ± .008 in (.2 mm). A company that made this film could control film thickness to ±.0008 in (.02 mm) consistently. The company made an economic decision to reduce the nominal thickness to .032 in (.82 mm), and with their ability to produce film within .0008 in of the nominal, all of the product would meet the national specification. This would reduce manufacturing costs and increase profits.

However, that same year strong typhoon winds caused a large number of the greenhouses to be destroyed. The cost to replace the film had to be

---

*Genichi Taguchi and Yu-in Wu, *Off-Line Quality Control*. Central Japan Quality Control Association, Nagaya, 1979, p. 7-8.

paid by the customer, and these costs were much higher than expected. Both of these cost situations can be seen in Fig. 1-4. What had not been considered by the film producer was the fact that the customer's cost would rise while the producer's cost was falling. The loss function, loss to society, is the upper curve, which is the sum of the producer's and customer's curves. This curve does show the proper thickness for the film to minimize loss to society, and this is where the nominal value of .039 in is located.

Looking at the loss function one can easily see that as the film gets thicker from the nominal of .039 in the producer is losing money, and when the film gets thinner from the nominal, the customer is losing money. The producer is obligated by being part of society to fabricate film with a nominal of .039 in and to reduce variation of that thickness to a low amount. In addition, it will save money for society to further reduce the manufacturer's capability from ±.0008 in (losses are lower closer to the nominal value).

If the producer does not attempt to hold the nominal thickness at .039 in and causes additional loss to society, then this is worse than stealing from the customer. If someone steals $10.00, the net loss to society is zero; someone has a $10.00 loss and the thief has a $10.00 gain. If, however, the producer causes an additional loss to society, everyone in society has suffered some loss. A producer who saves less money than the customer spends on repairs has done something worse than stealing from the customer. After this experience, the national specification was

Figure 1-4   Costs associated with greenhouse film. (*Reproduced from Genichi Taguchi and Yu-in Wu,* Off-line Quality Control. *Central Japan Quality Control Association, Nagaya, 1979, p. 7. Used by permission.*)

changed to make the average thickness produced .039 in. The tolerance was left unchanged at ±.0008 in.

1-6   Japan's Desire for Low Loss

One must understand Japan's position in worldwide economic competition to see how this philosophy was established. Japan is an island with limited natural resources except for a large group of people. The only method by which Japan can survive economically is to import materials, add value to the materials by processing, and export products. Much of Japan's economic success has been based on high-value-added products. High technology products, such as computers, are a natural for this approach. Success in the business of converting resources from one state (raw) to another state (finished) is due in part to the efficiency of this process. Efficiency of adding value to materials is equivalent to a low loss in the process. Low loss to society fits neatly within the niche of competing effectively in a value-added product arena.

1-7   Case Study: Tool Wear in a Process

This case study is intended to show a cost-oriented approach to quality control. With a very capable process, the goalpost philosophy allows tool wear to produce parts which vary from one specification limit to the other. The loss function economically justifies a different quality control approach. To illustrate this comparison, a new concept must first be introduced, a frequency, or probability of occurrence, distribution of products according to a measured value of a performance characteristic.

A typical frequency distribution for the heights of NBA basketball players might look like the histogram in Fig. 1-5. This graph says the height (performance characteristic of interest) of 80 in (2.03 m) is much more likely to occur than any other height, and this likelihood falls off as height goes higher or lower. This shape of distribution is known as a normal distribution, and is typically illustrated in the manner shown. Frequency distributions do not have to be bell-shaped as is this normal distribution, but many manufacturing processes and many natural phenomena produce results in this manner.

An important aspect of product and process quality is the relative widths of the distribution and the specification limits of the product. If the distribution is narrower than the specification limits, then it is possible to make all or nearly all of the parts to match the printed specification. This is especially true if the distribution is centered within the specification limits. All of this is very desirable from the goalpost point of view; however, what about the loss function viewpoint? An example will demonstrate the application of the loss function.

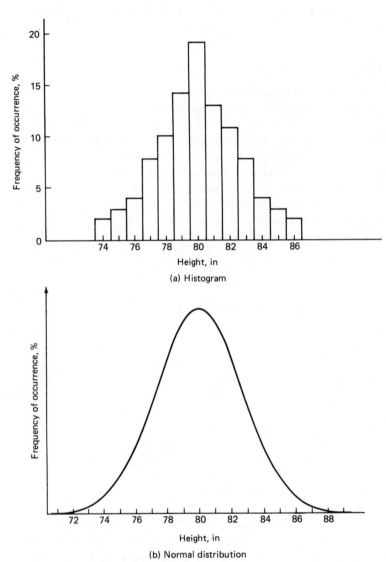

Figure 1-5 Frequency distributions of heights of NBA basketball players. (*a*) Histogram, (*b*) normal distribution.

A typical tolerance for a machined part, let's say an outer diameter, might be ±.010 in (±.25 mm). A process that would be very capable with respect to this tolerance would be modeled by the group of parts shown in Fig. 1-6. This histogram is for 100 separate parts.

The Taguchi loss function quantifies the variability present in a process. If a part reaches the end of the manufacturing line with a diameter exceeding the upper or lower limit, the part should be scrapped at a cost assumed to be $4.00. The scrap cost is only one aspect of loss to society. Presumably, the specifications are related to the reliability of the product; as the specification limits are approached, the product is less likely to provide satisfaction to the customer. If the product fails to perform satisfactorily, then other losses are incurred by the manufacturer or the customer, which makes scrap loss a conservative (low) estimate of loss to society.

Using this cost as a reference value, a loss function can be constructed for this situation as shown in Fig. 1-7. Taguchi uses the mathematical equation to model this picture of cost versus outer diameter:

$$L = k(y - m)^2 \qquad (1\text{-}1)$$

In this equation, L is the loss associated with a particular diameter value y, m is the nominal value of the specification, and the value of k is a constant depending on the cost at the specification limits and the width of the specification. In this example

$$L = k(y - m)^2$$

$$\$4.00 = k(\text{LSL} - m)^2$$

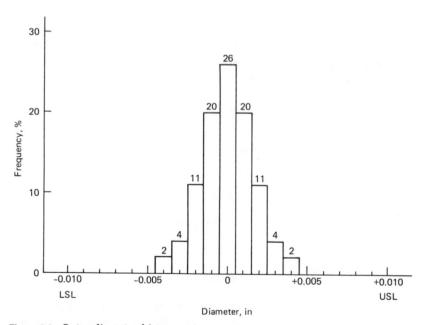

Figure 1-6 Outer diameter histogram.

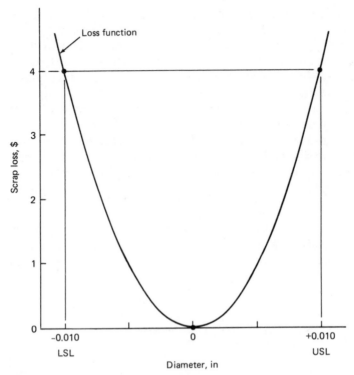

Figure 1-7 Outer diameter loss function.

The lower specification limit (LSL) is substituted into the equation, which is where the $4.00 loss is incurred. The upper specification limit also could be used for this calculation. Solving for k,

$$k = \frac{\$4.00}{(\text{LSL} - m)^2}$$

Given that $m = 0.0$ in (nominal value),

$$k = \frac{\$4.00}{(-.010 - 0.0)^2}$$

$$k = \$40,000 \text{ per in}^2$$

Therefore,

$$L = 40,000 \, (y - 0.0)^2 \qquad\qquad (1\text{-}2)$$

Now the loss associated with any part can be computed depending on the value of its diameter. For instance, a part with a diameter of +.003 in (.08 mm) costs

$$L = 40,000 \, (.003 - 0.0)^2 = \$.36$$

This is the loss per unit for each part shipped with an outer diameter of +.003 in.

The average cost per part for a particular group of parts also can be determined from this loss function. This can be accomplished in two ways. One way is to use the histogram of outer diameters and calculate the total cost for each value in the histogram, add the costs for each value, and divide by the total number of parts in the histogram. Using the dimensions for the parts shown in Fig. 1-6, one can calculate the loss. The numbers at the top of each bar indicate the quantity of parts with a particular outer diameter. By using the loss function, Eq. (1-2), the cost of a part of a particular diameter can be calculated.

$$L(-.002) = \$40,000 \, (-.002 - 0.0)^2 = \$.16$$

Since there are 11 parts having this value, the loss for all the parts having an outer diameter of −.002 in is $1.76. If we calculate this cost for each diameter, then the results are as shown in Table 1-1. The average loss per part is then

$$L = \frac{\$10.56}{100 \text{ parts}} = \$.11 \text{ per part}$$

A second method of estimating average loss per part entails using the loss equation in a slightly modified form. Mathematically, this calculation is equivalent to using the average value of the $(y - m)^2$ portion of the loss equation. Expanded, this is

TABLE 1-1   Cost of Parts versus Outer Diameter

Diameter, in	Loss/part, $	Total parts	Total loss, $
−.005	1.00	0	0.00
−.004	.64	2	1.28
−.003	.36	4	1.44
−.002	.16	11	1.76
−.001	.04	20	.80
.000	.00	26	0.00
.001	.04	20	.80
.002	.16	11	1.76
.003	.36	4	1.44
.004	.64	2	1.28
.005	1.00	0	0.00
	Grand totals	100	$10.56

$$L = k \, \frac{[(y_1 - m)^2 + (y_2 - m)^2 + \cdots + (y_N - m)^2]}{N} \qquad (1\text{-}3)$$

where N = number of parts sampled

If all the $(y - m)$ values are squared, added together, and divided by the number of items, then the result is the desired value. For a large number of parts, the average loss per part is equivalent to

$$L = k \, [S^2 + (\bar{y} - m)^2] \qquad (1\text{-}4)$$

S^2 = variance around the average, $\bar{y}$

$\bar{y}$ = average value of y for the group

$(\bar{y} - m)$ = offset of the group average from the nominal value m

Mathematically, Eqs. (1-3) and (1-4) are equivalent. A proof of the above situation is located in Appendix E.

For the above group of parts, the values of S^2 and $\bar{y}$ can be calculated using any statistical calculator and are

$$S^2 = 2.67 \times 10^{-6} \text{ in}^2 \ (1.72 \times 10^{-3} \text{ mm}^2)$$

$$\bar{y} = 0.0$$

$$L = k[S^2 + (\bar{y} - m)^2]$$

$$L = 40{,}000 \, [2.67 \times 10^{-6} + (0.0 - 0.0)^2]$$

$$L_1 = \$.11 \text{ per part}$$

This second method uses the general loss function for a nominal-is-best situation.

This example demonstrates the loss associated with a distribution that has a very low process capability ratio and is centrally located on the target value of 0.0 in. What if this machined feature causes the machining tool to wear, which subsequently causes an increasing diameter on sequential parts? Traditional quality control thinking (goalpost) would allow the situation shown in Fig. 1-8. The machine would be adjusted to the low side and allowed to wear until the high side was reached. Again, very few if any parts are outside the specification limits. What is the loss associated with each part on this basis?

The loss function uses the average and variance (standard deviation squared) for a group of parts to calculate loss. For a group of parts where the distribution average is continually increasing, the variance is

$$S^2_{\text{Resultant}} = S^2_{\text{Original}} + \frac{W^2}{12}$$

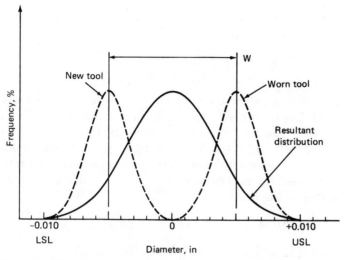

Figure 1-8 Distribution due to tool wear.

where W is the width of tool wear allowed. In this example,

$$S_O^2 = 2.67 \times 10^{-6} \text{ in}^2$$

$$W = .010 \text{ in}$$

$$S_R^2 = 2.67 \times 10^{-6} + \frac{.010^2}{12} = 1.10 \times 10^{-5}$$

The loss per part is then

$$L = k\,[S^2 + (\bar{y} - m)^2]$$

$$L_2 = \$40{,}000\,[1.10 \times 10^{-5} + (0 - 0)^2] = \$.44$$

One can readily see that the loss of a centered distribution is less than the loss of a distribution which traverses the specification range.

$$L_1 < L_2$$

But to obtain the low loss as in the first situation the machine would have to be adjusted after each part. For the second situation, the machine would not have to be adjusted nearly as often. There must, therefore, be some economic compromise between the loss function and adjustment costs.

The additional loss caused by situation 2 is

$$L_{\text{Additional}} = L_2 - L_1$$

$$L_A = k \left(S^2 + \frac{W^2}{12} \right) - k \, (S^2)$$

$$L_A = k \left(\frac{W^2}{12} \right)$$

The additional loss is due only to allowing the process to traverse across the tolerance band. The larger W becomes, the larger the loss. However, the larger W becomes, the less frequent the machine must be adjusted and the lower the adjustment costs. A plot of these two cost functions is shown in Fig. 1-9. The lowest total loss (optimum) is near the intersection of the two curves. The minimum value of $[L_A +$ (adjustment cost per part)] is located at the optimum adjustment width W.

C_A = adjustment cost per part for N parts

C_A = total machine adjustment cost divided by N

N = number of parts made since last adjustment

$N = W/R$

R = wear rate of tool (wear per part)

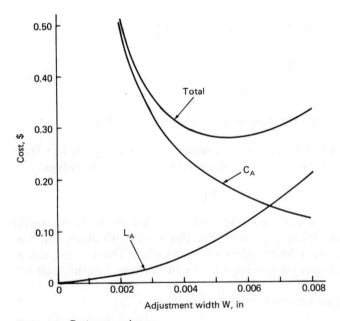

Figure 1-9   Cost comparison.

Substituting,

$$C_A = R \times \frac{\text{(total machine adjustment cost)}}{W}$$

Table 1-2 shows the optimum adjustment width for this example if these values are assumed

$$k = 40,000$$

$$R = .0005 \text{ in per part}$$

$$\$2.00 = \text{total machine adjustment cost}$$

The true optimum loss is at .005 in total wear span, but the additional loss of going up to .006 in total span is very minimal ($.003), which is beyond the accuracy of the loss function. Therefore, with a tool wear rate of .0005 in per part and an adjustment width of .006 in, the machine should be adjusted after 12 parts are produced. The average value for the outer diameter should be −.003 in after adjustment, and the tool should be allowed to wear until the average value for the outer diameter becomes +.003 in, providing a tool wear span of .006 in centered on the nominal value of 0.0 in.

Recall that the optimum adjustment interval depended on the scrap loss of one part, which was a conservative estimate of societal loss for the specification limit. If a larger loss is attributed to a part at the specification limit, the k value is increased, the loss function cost is increased, and the adjustment interval is subsequently decreased. When greater losses are incurred, greater uniformity of the product is necessitated.

Another contradiction also exists between the goalpost and loss function approach in this situation. If the process becomes more capable than depicted here, the goalpost approach would allow a wider tool wear span W. This in turn would increase the average loss per part associated with the resultant distribution. The loss function approach leaves the adjust-

TABLE 1-2 Adjustment Interval Calculation

W, in	L_A, $	C_A, $	Total, $
.002	.013	.500	.513
.003	.030	.330	.360
.004	.053	.250	.303
.005	.083	.200	.283
.006	.120	.166	.286
.007	.163	.143	.306
.008	.213	.125	.338

ment interval as before, since the additional loss due to wear is a function of W only, not variance. If the cost of adjustment were reduced, the goalpost approach would not require a change of the adjustment interval, whereas the loss function approach would change the adjustment interval to reduce the variation of the resultant distribution if the cost of adjustment were reduced.

This case study shows the economic penalty of allowing excess variation in a product or process. Taguchi uses a very comprehensive economic approach to on-line (production) quality control. A better description of the Taguchi methods would really be cost quality control.

1-8 Case Study: Automatic Transmissions*

Another example of the economic impact of excess variation occurred in the automatic transmission business with a major U.S. automobile company. Ford had contracted a Japanese supplier, Mazda, to make a certain portion of their front-wheel-drive automatic transmissions, with the balance of production made at a U.S. plant in Batavia, Ohio. Both sites were making transmissions to the same set of blueprints and the transmissions were being installed only in American cars. Mazda's version, as warranty records showed, had a substantially lower claim rate than the Batavia version. Ford investigated this phenomenon and found that Mazda's transmissions were made much more consistently than their own. On some critical control valve body components (valves, valve bores, and springs) which make a transmission shift automatically, Mazda was using only 27% of the allowed tolerance range, while Batavia was using 70%. Ford thought its plant was doing well, and by traditional standards it was, but the Mazda plant was superior. Not only were all the parts made to print, as were the U.S.-made parts, but they were more nearly like one another.

More investigation into the processes showed that Mazda was using a slightly more expensive and more complex grinder to finish the valve outer diameters. At first glance, one may think their parts were more expensive, but knowing that the loss function was at work, the parts were actually cheaper. The lower warranty bill for those transmissions substantiated that fact. By using this information, the Batavia plant has been able to improve its quality substantially and, in the first quarter of 1987, surpassed the Mazda level.

To summarize, continuous reduction of variation even within the allowed tolerance limits is a must to provide a more desirable product to the customer. This, in turn, is a more competitive product with a lower loss to society associated with it.

---

*Ford Motor Company, Dearborn, Michigan, 1987.

1-9 Factory Tolerances

One attribute of the loss function is to help determine what factory tolerances should be. For example, in automatic transmissions the shift points are supposed to occur at a certain speed at a certain throttle position. Heavy-duty truck users are particularly sensitive to this characteristic. Let's say that it costs the producer $100.00 to adjust a valve body under warranty when a customer complains of the shift point. From the information the engineers have available, the average customer would ask for an adjustment if the shift point is off from the nominal by 40 rpm transmission output speed on the first to second gear shift. The loss function is then

$$\text{Loss} = k\,(y - m)^2$$

$$\$100 = k\,(40)^2$$
$$k = \$.0625/\text{rpm}^2$$

At the factory the adjustment can be made at a much lower cost, approximately $10.00, which is comprised of the labor expense for time required to adjust the shift and rerun the test. What should be the limits that define when an adjustment should be made at the factory?

$$\text{Loss} = k\,(y - m)^2$$

$$\$10.00 = .0625\,(y - m)^2$$

$$(y - m)^2 = \frac{\$10}{.0625} = 160$$

$$(y - m) = \sqrt{160} = \pm 13 \text{ rpm}$$

Therefore, if the transmission shift point is further than 13 rpm from the desired nominal, it is cheaper to adjust it at the factory than to wait for a complaint and subsequent adjustment under warranty. The cost is always lower to make it right at the factory than to find a poor quality part when the customer has it in hand. Better yet, it is always lower cost to make it right at the point of production than to have to rework within the factory at a later time.

1-10 Other Loss Functions

The loss function can also be applied to product characteristics other than the situation where the nominal value is the best value; where lower is better or higher is better, for instance. A good example of a lower-is-better characteristic is the waiting time for your order delivery at a fast-food restaurant. If the attendant tells you that it will be a moment for your hamburger to come up, then you sense some loss; the longer you have to wait, the larger the loss. Microfinish of a machined surface, friction loss, or wear are also examples of lower is better. Efficiency, ultimate strength, or fuel economy are examples of higher is better.

The loss function for a lower-is-better characteristic is shown in Fig. 1-10. The cost constant k can be calculated similarly to the nominal-is-best situation. There is some loss associated with a particular value of y. The loss can then be calculated for any value of y based on that value of k. This loss function is identical to the nominal-is-best situation when $m = 0$, which is the best value for a lower-is-better characteristic (no negative values). This equation takes the form

$$L = k\,[S^2 + (\bar{y})^2]$$

The loss function for a higher-is-better characteristic is also shown in Fig. 1-10. Again, the cost constant can be calculated based upon some loss associated with a particular value of y. Subsequently, any value of y will have a loss assessed. The average loss per unit may be determined by finding the average value for $1/y^2$. This is mathematically equivalent to

$$L = k\left(\frac{1}{\bar{y}^2}\right)\left[1 + \left(\frac{3S^2}{\bar{y}^2}\right)\right]$$

Table 1-3 summarizes the different loss functions for the three types of characteristics, both for an individual part and for the average loss for a part from a distribution of parts.

1-11   General Loss Function for Nominal-Is-Best Situation

Returning once again to the general loss function for a nominal-is-best situation,

$$\text{Loss} = k\,[S^2 + (\bar{y} - m)^2]$$

One can see the equation is made up of two parts: the variance and the relative location of the average of a performance characteristic of a group of products. Therefore, to minimize loss to society, the product characteristic needs to be centered at the nominal value and the variance of that characteristic needs to be reduced.

The loss function entails two aspects of quality management within a factory. One, the variance, is the product and process engineer's job to establish before the start of production and to improve as time goes on (lower loss). And two, the centering of the distribution is the responsibility of the production (manufacturing) people on a day-in, day-out

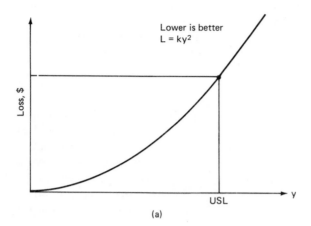

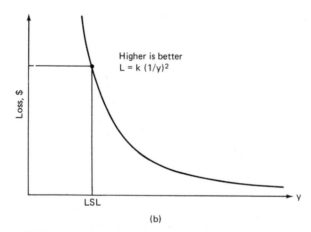

Figure 1-10   Other loss functions. (a) Lower is better, $L = ky^2$; (b) higher is better, $L = k(1/y^2)$.

TABLE 1-3 Types of Loss Functions

Type of characteristic	Loss for an individual part	Average loss per part in a distribution
Higher is better	$k(1/y^2)$	$k[1/\bar{y}^2][1 + (3S^2/\bar{y}^2)]$
Nominal is best	$k(y - m)^2$	$k[S^2 + (\bar{y} - m)^2]$
Lower is better	$k(y^2)$	$k[S^2 + (\bar{y}^2)]$

basis. This is off-line quality control (designers) and on-line quality control (makers), with reference to Fig. 1-1. The loss function is not intended to be accurate to the hundredth of a penny but is intended to show how reduced variation can result in reduced losses.

1-12   Summary

This chapter offered a different view of quality than has traditionally been used; uniformity of products is of greater concern than just conformance to specifications. Also considered was the economic impact of reducing variation in products and processes. Several examples were offered to show the relationship of variability to customer desires and a way of quantifying the value (loss function) of quality (reduced variation). The loss function addressed two descriptive statistics of a group of products: the average and the variance. The means to achieve a certain average from a product or process will not be addressed, since this is manufacturing's responsibility. Continuous reduction in the variance, however, can be intentionally achieved through the use of the methods covered in the remaining chapters of this text.

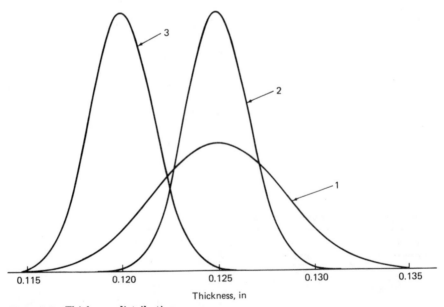

Thickness, in

Figure 1-11 Thickness distributions.

Problems

1-1   Three processes make frictional clutch plates for an automatic transmission; none of the processes makes more than .27% scrap relative to the upper and lower specification limits. The drawing specifies an overall thickness of .125 in ± .010 in. The cost to scrap is $2.50 per part. What is the k value in the loss function?

1-2   What is the average loss per part for each of the groups of parts shown in Fig. 1-11? The population limits shown are assumed to be ±3 sigma limits.

1-3   What conclusions are there concerning variation and the location of the average of the populations?

Problems

Introduction to Analysis of Variance

2-1 Understanding Variation

The purpose of product or process development is to improve the performance characteristics of the product or process relative to customer needs and expectations. The purpose of experimentation should be to reduce and control variation of a product or process; subsequently, decisions must be made concerning which parameters affect the performance of a product or process. The loss function quantifies the need to understand which design factors influence the average and variation of a performance characteristic of a product or process. By properly adjusting the average and reducing variation, the product or process losses are minimized.

Since variation is a large part of the discussion relative to quality, analysis of variance (ANOVA) will be the statistical method used to interpret experimental data and make the necessary decisions. This method was developed by Sir Ronald Fisher in the 1930s as a way to interpret the results from agricultural experiments. ANOVA is not a complicated method and has a lot of mathematical beauty associated with it. ANOVA is a statistically based decision tool for detecting any differences in average performance of groups of items tested. The decision, rather than using pure judgment, takes variation into account.

The discussion of ANOVA will start with a very simple case and build up to more comprehensive situations in this chapter. In the next chapter, ANOVA will be applied to some very specialized experimental situations, although this analysis method can be used with any set of data that has some structure. The experimental designs and subsequent analyses are intrinsically tied to one another, but the understanding of the experimental designs will be much easier if the groundwork for the analytical method is laid first.

2-2 No-Way ANOVA

No traditional statistics book will mention no-way ANOVA, but it is the simplest situation to analyze and has some practical value which will be discussed in later chapters.

2-2-1 Sums of squares

No-way ANOVA begins with a set of experimental data, real or, as in this case, assumed for simplification. Imagine an engineer is sent to a production line to sample a set of windshield washer pumps for the purpose of measuring flow rate. The data collected could be like that in Table 2-1.

Although this data is entirely fictitious and put into single digits to simplify calculations, these observations are quite possible. ANOVA would mathematically be carried out in an identical manner regardless of the data actually collected or the units of the data points. Many of the examples in this book are simplified in this manner to demonstrate the analysis method rather than to test the mathematical ability of the reader.

Analysis of variance is a mathematical technique which breaks total variation down into accountable sources; total variation is decomposed into its appropriate components. No-way ANOVA, the simplest case, breaks total variation down into only two components:

1. The variation of the average (or mean) of all the data points relative to zero

TABLE 2-1 Low-Capacity Pump Flow Rate

Pump no.	1	2	3	4	5	6	7	8
Flow rate, oz/min	5	6	8	2	5	4	4	6

NOTE: 1 oz/min = .473 ml/s

2. The variation of the individual data points around the average (traditionally called experimental error)

Some notation is necessary to demonstrate the calculation method:

$$y = \text{observation, response, data}$$

$$y_i = i\text{th response; } y_3 = 8 \text{ oz/min}$$

$$N = \text{total number of observations}$$

$$T = \text{sum of all observations}$$

$$\overline{T} = \text{average of all observations} = T/N = \overline{y}$$

In this case

$$N = 8 \qquad T = 40 \text{ oz/min} \qquad \overline{T} = 5.0 \text{ oz/min}$$

The word "observation" has a particular connotation. The true flow rate of any given pump is actually unknown; it is only estimated through the use of some flow meter. There will be some unknown measurement error present, but a flow rate will nonetheless be observed and accepted as the pump's performance under the conditions of the test. Also, these pumps were randomly selected from a production line which ostensibly manufactures identical pumps; however, there will be slight differences from pump to pump, causing a portion of the pump-to-pump variation in performance. Why are not all of the pump flow rates identical; would we expect them to be?

No-way ANOVA can be illustrated graphically, which enhances the reader's opportunity to understand. A plot of the data appears in Fig. 2-1. The magnitude of each observation can be represented by a line segment extending from zero to the observation. These line segments can be divided into two portions: one portion attributed to the mean and one portion attributed to error. Error includes the measurement errors plus

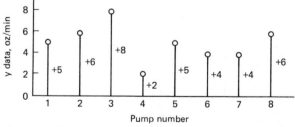

Figure 2-1 Low-capacity pump performance; total variation.

the effects of pump characteristics that are varying slightly from pump to pump.

The magnitude of the line segment due to the mean is indicated by extending a line from the average value to zero as shown in Fig. 2-2. The magnitude of the line segment due to error is indicated by the difference of the average value from each observation as shown in Fig. 2-3.

Note the summation of the line segments due to the mean equals 40.0, which is also equal to the sum of the raw data points. The summation of the error term around the average is equal to zero. This in itself is not very informative; however, there is a mathematical operation to be performed which allows a clearer picture to develop. The magnitudes of each of the line segments can be squared and then summed to provide a measure of the total variation present. Referring to Fig. 2-1, the total sums of squares (total variation) is then computed:

$$SS_T = \text{total sums of squares}$$

$$SS_T = 5^2 + 6^2 + 8^2 + 2^2 + 5^2 + 4^2 + 4^2 + 6^2$$

$$SS_T = 222.0$$

The magnitude of the portion of the line segment due to the mean can also be squared and summed. Referring to Fig. 2-2, the variation due to the mean is then computed:

$$SS_m = \text{sums of squares due to the mean}$$

$$SS_m = N(\overline{T})^2$$

But $\overline{T} = T/N$, so

$$SS_m = N\left(\frac{T}{N}\right)^2 = \frac{T^2}{N}$$

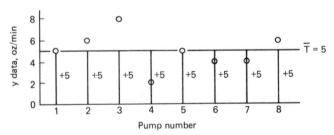

Figure 2-2 Low-capacity pump performance; variation due to mean.

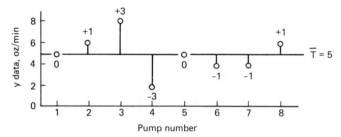

Figure 2-3   Low-capacity pump performance; variation due to error.

$$SS_m = \frac{40^2}{8} = 200.0$$

The portion of the magnitude of the line segment due to error can be squared and summed to provide a measure of the variation around the average value. Referring to Fig. 2-3, the variation due to error is

SS_e = error sums of squares

$SS_e = 0^2 + 1^2 + 3^2 + (-3)^2 + 0^2 + (-1)^2 + (-1)^2 + 1^2$

$SS_e = 22.0$

Note that

$$222.0 = 200.0 + 22.0$$

This demonstrates (not a proof) a basic property of ANOVA. The total sums of squares is equal to the sum of the sums of squares due to the known components. In this case,

$$SS_T = SS_m + SS_e$$

In this method the total variation can be decomposed into two sources with the appropriate share apportioned to each source. The formulas for the sums of squares can be written generally

$$SS_T = \sum_{i=1}^{N} y_i^2$$

which is the summation of the squares of each observation from $i = 1$ to N, and

$$SS_m = \frac{T^2}{N}$$

which is equivalent to the summation of the square of the portion of each observation due to the mean for $i = 1$ to N.

$$SS_e = \sum_{i=1}^{N} (y_i - \overline{T})^2$$

which is the summation of the square of the differences of each observation from the mean.

Here, the error component was actually calculated, but this was not really necessary. The ANOVA method used states that

$$SS_T = SS_m + SS_e$$

so

$$SS_e = SS_T - SS_m = 222 - 200 = 22$$

Practice problem 1. Another model of windshield washer pump is sampled from production (see Table 2-2). It is recommended that the reader perform a no-way ANOVA on this set of observations. Answers are in Appendix A.

2-2-2 Degrees of freedom

To complete the ANOVA calculations, one other element must be considered, that being degrees of freedom. A degree of freedom in a statistical sense is associated with each piece of information that is estimated from the data. For instance, the mean (average) is estimated from all the data and requires one degree of freedom (d.f.) for that purpose. Another way to think of the concept of degrees of freedom is to allow 1 d.f. for each fair (independent) comparison that can be made in the data. Only one fair comparison can be made between the mean of all the data (there is only one mean) and zero, the reference point; 5.0 oz/min only has meaning when thought of in terms of the reference point 0.0 oz/min. Therefore, there is only 1 d.f. associated with the mean, which is only one piece of information.

This concept of independent comparisons also applies to the degrees of freedom associated with the error estimate. With reference to the eight observations, there are seven independent comparisons that can be made to estimate the variation in the data. Data point 1 can be compared to data point 2, 2 to 3, 3 to 4, etc., as shown in Fig. 2-4. In this data set there exist only seven independent comparisons. Data point 1 compared

TABLE 2-2 High-Capacity Pump Flow Rate

Pump no.	1	2	3	4	5	6	7	8
Flow rate, oz/min	15	16	18	12	15	14	14	16

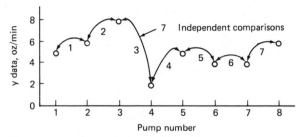

Figure 2-4 Low-capacity pump performance; degrees of freedom.

to data point 3 is not another independent comparison; that comparison is dependent on the comparison of data point 1 to 2 and 2 to 3.

Similar to sums of squares, a summation can be made for degrees of freedom, where

$$SS_T = SS_m + SS_e$$

let ν = degrees of freedom

ν_T = total degrees of freedom

ν_m = degree of freedom associated with the mean (always 1)

ν_e = degrees of freedom associated with error

$$\nu_T = \nu_m + \nu_e$$

$$8 = 1 + 7$$

The total degrees of freedom equals the total number of observations in the data set for this method of ANOVA.

An ANOVA table can now be constructed which summarizes all the information for the low-capacity pump data set (see Table 2-3).

2-2-3   Variance due to error

One other descriptive statistic that can be calculated from the ANOVA table is the variance V. Error variance, usually termed as just variance, is equal to the sums of squares for error divided by the degrees of freedom for error; in the example,

$$V_e = \frac{SS_e}{\nu_e}$$

$$V_e = \frac{22}{7} = 3.14$$

TABLE 2-3 No-Way
ANOVA Summary

Source	SS	d.f.
Mean	200	1
Error	22	7
Total	222	8

By definition, the standard deviation is equal to the square root of variance.

$$S = \sqrt{V}$$

S = sample standard deviation

σ = population standard deviation

S is an estimate of σ, the true but unknown standard deviation.

This calculation is identical to the standard deviation calculation done by an electronic calculator. The formula used is

$$S = \sqrt{\frac{\sum\limits_{i=1}^{N}(y_i - \bar{y})^2}{N - 1}} \qquad \text{standard deviation}$$

$$S^2 = V = \frac{\sum\limits_{i=1}^{N}(y_i - \bar{y})^2}{N - 1} \qquad \text{variance}$$

Notice that the numerator is the same calculation as for the sums of squares due to error, and that the denominator is the degrees of freedom associated with error variation. Although the formula above is much faster than ANOVA for calculating error variance in the example given, when the experimental situations become more complex, ANOVA will become the faster method.

Error variance is a measure of the variation due to all the uncontrolled parameters, including measurement error involved in a particular experiment (set of data collected).

2-2-4 Method of least squares

This ANOVA method is based on a least squares approach; the error variance is equal to the minimum value of the sums of squares about some reference value divided by the degrees of freedom for error. Notice that the practice problem had exactly the same error variance as the example problem, even though the zero reference line was 10 units further away (a value of 10 was added to all the original observations). It so happens that the error sums of squares are minimized when the reference line is the average value of the observations.

2-3   One-Way ANOVA

One-way ANOVA is the next most complex ANOVA to conduct. This situation considers the effect of one controlled parameter upon the performance of a product or process, in contrast to no-way ANOVA, where no parameters were controlled.

2-3-1   Product development situation

Again, an imaginary, yet potentially very real, situation is presented as a typical case when one-way ANOVA is applicable. Imagine the same engineer as before is charged with the task of establishing the fluid velocity generated by the windshield washer pumps. Obviously, if the fluid velocity is too low, the fluid will merely dribble out, and if too high, air movement past the windshield will not be able to distribute the cleaning fluid adequately to satisfy the driver of the car. The engineer proposes a test of three different orifice areas to determine which may give a proper fluid velocity.

Before the test data is collected some notation is in order to simplify the mathematical discussion.

$$A = \text{factor under investigation (outlet orifice area)}$$

$$A_1 = \text{1st level of orifice area} = .0015 \text{ in}^2 \ (.97 \text{ mm}^2)$$

$$A_2 = \text{2nd level of orifice area} = .0030 \text{ in}^2 \ (1.94 \text{ mm}^2)$$

$$A_3 = \text{3rd level of orifice area} = .0045 \text{ in}^2 \ (2.90 \text{ mm}^2)$$

The same symbol for the level designation will be used to denote the sum of responses for that test condition.

$$A_i = \text{sum of observations under } A_i \text{ level}$$

$$\overline{A}_i = \text{average of observations under } A_i \text{ level} = A_i/n_{A_i}$$

$$T = \text{sum of all observations}$$

$$\overline{T} = \text{average of all observations} = T/N$$

$$n_{A_i} = \text{number of observations under } A_i \text{ level}$$

$$N = \text{total number of observations}$$

$$k_A = \text{number of levels of factor } A$$

With this notation in mind, the engineer constructs four pumps with a given orifice area, making a total of 12 to test. The test data is shown in Table 2-4. Using the notation system, A_1 represents .0015 in² orifice area and also the sum of observations of the pump with .0015 in² orifice area, which is 8.8 ft/s. Then,

$$A_1 = 8.8 \text{ ft/s} \qquad n_{A_1} = 4 \qquad \overline{A}_1 = 2.2 \text{ ft/s}$$

$$A_2 = 5.1 \text{ ft/s} \qquad n_{A_2} = 3 \qquad \overline{A}_2 = 1.7 \text{ ft/s}$$

$$A_3 = 3.2 \text{ ft/s} \qquad n_{A_3} = 4 \qquad \overline{A}_3 = .8 \text{ ft/s}$$

$$T = 17.1 \text{ ft/s} \qquad N = 11 \qquad \overline{T} = 1.6 \text{ ft/s}$$

$$k_A = 3$$

2-3-2   Sums of squares (one-way ANOVA)

Two methods can be used in ANOVA to complete the calculations:

1. Including the mean (Sec. 2-3-3)
2. Excluding the mean (Sec. 2-3-6)

2-3-3   Method 1 (including the mean)

As before, the total variation can be decomposed into its appropriate components.

1. The variation of the average (mean) of all observations relative to zero
2. The variation of the average (mean) of observations under each factor level around the average of all observations
3. The variation of the individual observations around the average of observations under each factor level

The calculations are identical to the no-way ANOVA example, with

TABLE 2-4   Pump Velocity Data (First Experiment)

Level	Area, in²	Velocity, ft/s				Total
A_1	.0015	2.2	1.9	2.7	2.0	8.8
A_2	.0030	1.5	1.9	1.7	*	5.1
A_3	.0045	.6	.7	1.1	.8	3.2
					Grand total	17.1

*Dropped pump and destroyed it, no data.
NOTE: 1 ft/s = .30 m/s.

the exception of the component of variation due to factor A, outlet orifice area. Graphically, this can be seen quite clearly in Fig. 2-5.

$$SS_T = \sum_{i=1}^{N} y_i^2$$

$$SS_T = 2.2^2 + 1.9^2 + 2.7^2 + \cdots + .8^2 = 31.190$$

Figure 2-6 shows graphically the line segment which is proportional to the variation due to the mean.

$$SS_m = N\,(\overline{T})^2 = \frac{T^2}{N}$$

$$SS_m = \frac{17.1^2}{11} = 26.583$$

This calculation is based on the format used in previous sums of squares calculations. The magnitude of the line segments due to each level of

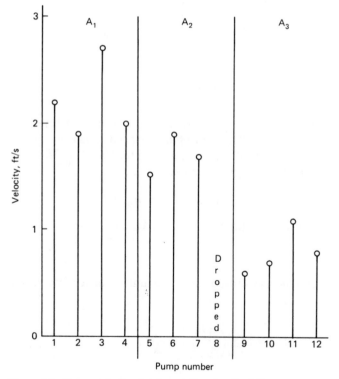

Figure 2-5 Pump velocity; total variation (method 1).

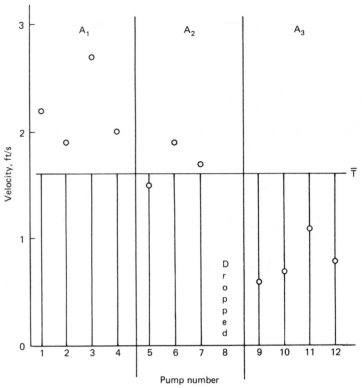

Figure 2-6 Pump velocity; variation due to mean.

factor A is squared and summed. For instance, the length of line segment due to level A_1 is $(A_1 - \overline{T})$. There are four observations under level A_1 condition. The same type of information is used for the other levels of factor A. Figure 2-7 shows this graphically.

$$SS_A = n_{A_1}(\overline{A}_1 - \overline{T})^2 + n_{A_2}(\overline{A}_2 - \overline{T})^2 + n_{A_3}(\overline{A}_3 - \overline{T})^2$$

$$SS_A = 4(.64545)^2 + 3(.14545)^2 + 4(-.75454)^2$$

$$SS_A = 4.007$$

This calculation method is very tedious, but it is mathematically equivalent to

$$SS_A = \left[\sum_{i=1}^{k_A} \left(\frac{A_i^2}{n_{A_i}} \right) \right] - \frac{T^2}{N}$$

(See Appendix E, proof 2.)

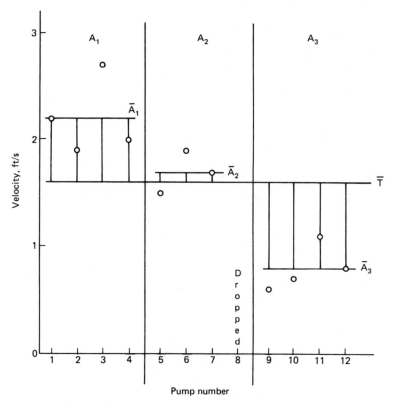

Figure 2-7 Pump velocity; variation due to factor A.

$$SS_A = \frac{8.8^2}{4} + \frac{5.1^2}{3} + \frac{3.2^2}{4} - \frac{17.1^2}{11} = 4.007$$

Figure 2-8 shows the remaining portion of the line segment, which is proportional to the variation due to error.

$$SS_e = \sum_{j=1}^{k_A} \sum_{i=1}^{n_{A_j}} (y_i - \overline{A}_j)^2$$

$$SS_e = 0^2 + (-.3)^2 + .5^2 + (-.2)^2 + (-.2)^2 + .2^2$$
$$+ 0^2 + (-.2)^2 + (-.1)^2 + .3^2 + 0^2 = .600$$

Error variation is again based on the method of least squares, but in one-way ANOVA the least squares are evaluated around the average for each level of the controlled factor. Error variation is the uncontrolled variation within the controlled groups. Again, the total variation can be accounted for mathematically.

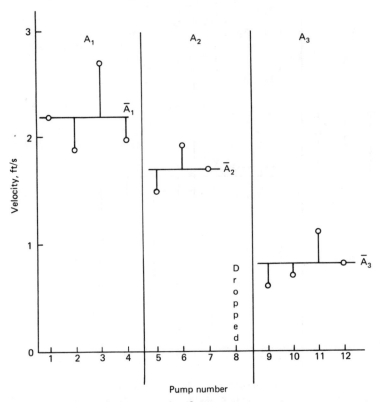

Figure 2-8 Pump velocity; variation due to error.

$$SS_T = SS_m + SS_A + SS_e$$

$$31.190 = 26.583 + 4.007 + .600$$

2-3-4 Degrees of freedom
(including the mean)

In method 1, including the mean, the total degrees of freedom is the sum of degrees of freedom for the mean, for factor A, and for error.

$$v_T = v_m + v_A + v_e$$

$$v_T = N = 11$$

$$v_A = k_A - 1 = 3 - 1 = 2$$

$$v_e = v_T - v_m - v_A = 11 - 1 - 2 = 8$$

The degrees of freedom for a controlled factor such as A follow the same concept of 1 d.f. for each fair comparison. $\overline{A}_1$ compared to $\overline{A}_2$ has 1 d.f.

and $\overline{A}_2$ compared to $\overline{A}_3$ has 1 d.f. Also apparent in the data set are the eight degrees of freedom for error, eight fair comparisons within groups. Error estimation is always done within groups. For instance, the comparison of 1.5 to .6 is one estimate of the A_2 to A_3 effect, not an estimate of error. Table 2-5 shows the factor and error degrees of freedom.

2-3-5 ANOVA summary table (including the mean)

The ANOVA summary is shown in Table 2-6. Now, as well as error variance being estimated, the variance due to factor A is estimated. The meaning of this estimate and the relevance will be discussed later.

2-3-6 Method 2 (excluding the mean)

It has been shown in a previous example that the variation due to the mean does not affect the calculations for the variation due to error, and for edification, neither does it affect the calculations for the factor effects. In most experimental situations, with the exception of lower-is-better characteristics, the variation due to the mean has no practical value. In a lower-is-better situation the variation due to the mean is a measure of how far the average is from zero and how successful the factors might be in reducing the average to zero.

In the situation of the exclusion of the mean from the ANOVA calculations, total variation may be decomposed into

1. The variation of the average of observations under each factor level around the average of all observations

2. The variation of the individual observations around the average of observations under each factor level.

TABLE 2-5 Factor and Error Degrees of Freedom

Level	Velocity, ft/s	$\overline{A}$		
A_1	2.2 1.9 2.7 2.0	2.2		
	1 2 3		1	
A_2	1.5 1.9 1.7 *	1.7		2 d.f. for
	4 5			factor A
A_3	.6 .7 1.1 .8	.8	2	
	6 7 8			
	8 d.f. for error			

*Dropped pump and destroyed it, no data.

TABLE 2-6 One-Way ANOVA Summary (Method 1)

Source	SS	v	V
m	26.583	1	26.583
A	4.007	2	2.004
e	.600	8	.075
T	31.190	11	

Again, graphically this can be demonstrated as shown in Fig. 2-9. The same concept of summing the squares of the magnitudes of the various line segments is applied in method 2 also.

$$SS_T = \sum_{i=1}^{N} y_i^2$$

$$SS_T = .645^2 + .345^2 + 1.145^2 + \cdots + (-.755)^2$$

$$SS_T = 4.607$$

Mathematically, this is equivalent to

$$SS_T = \left[\sum_{i=1}^{N} y_i^2 \right] - \frac{T^2}{N} \tag{2-1}$$

(See Appendix section E, proof 3). From previous calculations,

$$SS_T = 31.190 - 26.583 = 4.607$$

Equation 2-1 will be used primarily as the definition for total variation. The variation due to factors is calculated identically to method 1; refer to Fig. 2-7.

$$SS_A = \left[\sum_{i=1}^{k_A} \left(\frac{A_i^2}{n_{A_i}} \right) \right] - \frac{T^2}{N}$$

The variation due to error is also identical to method 1; refer to Fig. 2-8.

$$SS_e = \sum_{j=1}^{k_A} \sum_{i=1}^{n_{A_j}} (y_i - \overline{A}_j)^2$$

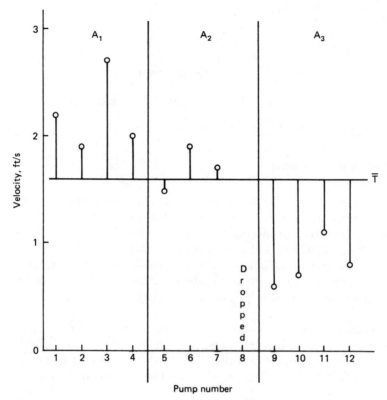

Figure 2-9 Pump velocity; total variation (method 2).

2-3-7 Degrees of freedom (excluding the mean)

In method 1, the degrees of freedom formula was written

$$v_T = v_m + v_A + v_e$$

$$v_m = 1 \text{ (always)}; \qquad v_T = N$$

In method 2, the mean being excluded, the degree of freedom for the mean is subtracted from both sides of the equation

$$N = 1 + v_A + v_e$$

$$N - 1 = v_A + v_e$$

A new definition used in method 2 is

$$v_T = N - 1$$

Therefore,

$$v_T = v_A + v_e$$

$$v_T = N - 1 = 11 - 1 = 10 \qquad \text{for the total d.f.}$$

$$v_A = k_A - 1 = 3 - 1 = 2 \qquad \text{for the d.f. for factor } A$$

$$v_e = v_T - v_A = 10 - 2 = 8 \qquad \text{for the d.f. for error}$$

2-3-8 ANOVA summary table (excluding the mean)

The ANOVA summary table for method 2 appears in Table 2-7. The identical variance values are calculated for factor A and error in one-way ANOVA using method 1 or method 2. The value for the mean is disregarded in method 2, which is the most popular method. Only in a case where the performance parameter is a lower-is-better characteristic would the variance due to the mean be relevant; this provides a measure of how effective some factor might be in reducing the average to zero.

Practice problem 2. Imagine a similar experiment on the higher-capacity pump using generally larger orifices with the results shown in Table 2-8. It is again recommended that the reader calculate the items necessary for the ANOVA table using method 2. The results are shown in Table 2-9.

2-3-9 Central limit theorem

The variance of factor A, shown in the ANOVA tables, is an estimate of error variance based upon the variation of the averages $\overline{A}_1$, $\overline{A}_2$, and $\overline{A}_3$ around the grand average $\overline{T}$. The error variance shown in the ANOVA tables is based upon the variation of the actual individual observations within a controlled group (the data points under condition A_1, for example). This situation is based on one of the fundamental theorems of statistics, the central limit theorem (CLT). The CLT has three tenets

TABLE 2-7 One-Way ANOVA Summary (Method 2)

Source	SS	v	V
A	4.007	2	2.004
e	.600	8	.075
T	4.607	10	

TABLE 2-8 Pump Velocity Data (Second Experiment)

Level	Area, in²		Velocity, ft/s			Total
A_1	.008	.8	1.2	.9	1.3	4.2
A_2	.006	1.5	1.4	1.2	1.8	5.9
A_3	.004	2.3	1.9	1.9	2.1	8.2
				Grand total		18.3

1. Sample averages tend to be normally distributed regardless of the distribution of the individuals
2. The average of the distribution of sample averages will approach the average of the distribution of the individuals
3. The variance of the sample averages is less than the variance of the distribution of the individuals

All of the above are predicated on the fact that one population is being sampled; therefore, it has a constant average and constant standard deviation.

The formula that describes the third and most important tenet is

$$\sigma_{\bar{y}}^2 = \frac{\sigma_y^2}{n} \qquad (2\text{-}2)$$

This formula states that the variance of sample averages will be equal to the variance of the individuals divided by the sample size used to obtain the sample averages. Graphically, this appears in Fig. 2-10. Formula (2-2) may be rewritten

$$\sigma_y^2 = n\sigma_{\bar{y}}^2$$

As estimates of σ, use S

$$S_y^2 = n \, (S_{\bar{y}})^2$$

TABLE 2-9 One-Way ANOVA Summary

Source	SS	ν	V
A	2.0150	2	1.008
e	.4675	9	.052
T	2.4825	11	

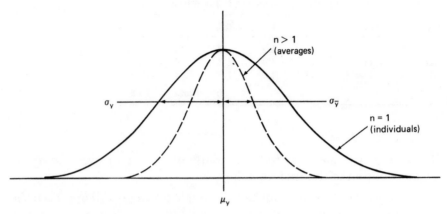

Figure 2-10 Central limit theorem.

2-3-10 Variance calculation using CLT

This formula can be used to make an estimate of individual variance by taking the variance of the sample averages and multiplying by the sample size (equal sample size for all samples is required). This property of the CLT can be applied to the practice problem, where all levels had a sample size of $n = 4$.

The variance of sample averages $\overline{A}_1$, $\overline{A}_2$, and $\overline{A}_3$ can be calculated using the variance formula modified for averages. Rather than this formula for individuals,

$$S_y^2 = \frac{\sum_{i=1}^{N}(y_i - \overline{T})^2}{N - 1}$$

Substitute k_A for N and $\overline{A}$ for y.

$$S_{\overline{A}}^2 = \frac{\sum_{i=1}^{k_A}(\overline{A}_i - \overline{T})^2}{k_A - 1}$$

$$S_{\overline{A}}^2 = \frac{(1.050 - 1.525)^2 + (1.475 - 1.525)^2 + (2.050 - 1.525)^2}{3 - 1}$$

$$S_{\overline{A}}^2 = .252$$

But

$$S_y^2 = n\,(S_{\overline{A}})^2 \qquad \text{where } n = 4$$

$$S_{y_1}^2 = 4\,(.252) = 1.008$$

Notice that this value is equal to the variance of factor A in the ANOVA table, but it is really an estimate of individual variance based on the variance of sample averages.

2-3-11   Individual Variance Calculation

Individual variance can be estimated one other way, based on the actual individual variation. Using the formula for individual variance, the variance within the A_1 group can be calculated:

For the A_1 group

$$S_{y_2}^2 = .0570 \qquad \text{where } n = 4 \text{ and } v = 3$$

For the A_2 group

$$S_{y_2}^2 = .0630 \qquad \text{where } n = 4 \text{ and } v = 3$$

For the A_3 group

$$S_{y_2}^2 = .0370 \qquad \text{where } n = 4 \text{ and } v = 3$$

Each of these is an estimate of individual variance with 3 d.f. Which is the correct value? Obviously, all cannot be correct. Since all of this data is available, an obligation exists to use all of the information rather than a part of the information. A better estimate uses all of the information by averaging them:

$$S_{y_2}^2 = \frac{.0570 + .0630 + .0370}{3} = .052$$

Now there are two independent estimates of individual variance; however, these estimates appear to be considerably different in this case:

$$1.008 \gg .052$$

$$S_{y_1}^2 \gg S_{y_2}^2$$

$S_{y_1}^2$ being determined from the variation of averages and $S_{y_2}^2$ being determined from the variation of individuals.

2-3-12   *F* test for variance comparison

Statistically, there is a tool which provides a decision at some confidence level as to whether these estimates are significantly different. This tool is called an *F* test, named after Sir Ronald Fisher, a British statistician, who invented the ANOVA method. The *F* test is simply a ratio of sample variances.

$$F = \frac{S_{y_1}^2}{S_{y_2}^2}$$

When this ratio becomes large enough, then the two sample variances are accepted as being unequal at some confidence level. F tables which list the required F ratios to achieve some confidence level are provided in the appendixes. To determine whether an F ratio of two sample variances is statistically large enough, three pieces of information are considered. One, the confidence level necessary; two, the degrees of freedom associated with the sample variance in the numerator; and three, the degrees of freedom associated with the sample variance in the denominator. Each combination of confidence, numerator degrees of freedom, and denominator degrees of freedom has an F ratio associated with it. $F_{\alpha;\nu_1;\nu_2}$ is the format for determining an explicit F value where

$$\alpha = \text{risk}$$

$$\text{Confidence} = 1 - \text{risk}$$

$$\nu_1 = \text{degrees of freedom for numerator}$$

$$\nu_2 = \text{degrees of freedom for denominator}$$

In this example, an assumed confidence level of 90% is required so the risk becomes 10%. The degrees of freedom for the numerator is 2 and for the denominator is 9. The necessary F ratio to look for in the tables is

$$F_{.10;2;9}$$

Looking in the appropriate table,

$$F_{.10;2;9} = 3.01$$

In the practice problem the ratio of the two estimates of individual variance is:

$$F = \frac{1.008}{.052} = 19.4$$

Comparing the F value of the data with the table,

$$F_{\text{data}} >> F_{.10;2;9}$$

Statistically, this means with at least 90% confidence the two estimates of variance are believed to be unequal. What does this mean from a practical viewpoint?

The estimate of individual variance based on variation of averages is inappropriately too high when considering the actual individual variance. Averages are varying much more than would be expected from the individual variation that is present. Rather than believing that only one population is being sampled, it is believed two or more having different averages are being sampled. The data from the practice problem in Sec. 2-3-9 is graphically displayed in Fig. 2-11. In another form the graph of the data would appear as in Fig. 2-12. In summary, at a very high confidence level (note: the F ratio from the practice problem exceeds the F ratio from the tables for 99% confidence level), the outlet orifice area affects flow velocity generated by the windshield washer pump.

As the F ratio is applied in the ANOVA case, the alpha risk is the chance of obtaining a ratio of at least the magnitude indicated in the table when pulling samples out of the same population (a group of items having the same mean and variance). Since the alpha risk is chosen as a small value, the experimenter would rather believe that two or more populations with different averages have been sampled when the F ratio does attain the specified magnitude. The odds are great against samples from one population attaining the F ratio; it is much more easily obtained when two or more populations are considered.

Confidence in the statistical sense, as it is used in this example, means there is some chance of error in the statement to which the confidence applies. For instance, the confidence that a number of 1, 2, 3, 4, or 5 could be rolled on a standard die is 5/6, or 83%. It is possible to roll a 6, so there is some risk of being wrong. When stating a confidence level for a factor, the experimenter is simply stacking the odds in their favor that the factor's levels really do have different average performances. A high confidence value may be chosen to reduce the risk of an error, alpha, but a larger actual difference in performance will be required to make the risk lower. In this situation the alpha risk is the chance that a factor really does not cause a difference in performance when the experi-

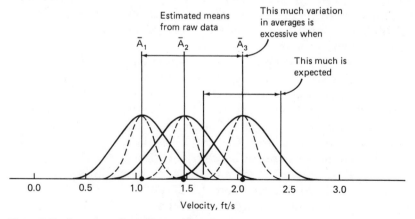

Figure 2-11 Averages of practice problem 2.

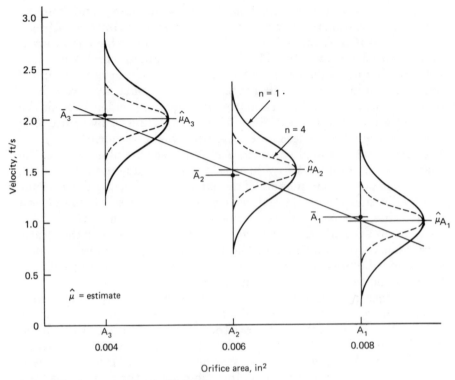

Figure 2-12 Averages versus orifice area.

menter, based on the data at hand, believes that it does. The magnitude of the F ratio does not necessarily reflect the engineering importance of the factorial effect. There could exist a very high F ratio, if error variance were very small relative to the factor variance, with only a small difference (from an engineering viewpoint) in level averages. The confidence may be high, but the difference in level averages may not be practically useful to the experimenter.

The performance of each pump is best estimated to be linear over the range of orifice areas tested because the sample averages are within the expected distance of a set of averages falling on a straight line. A method to statistically evaluate nonlinear performance will be covered later in Chap. 4. To obtain some particular flow velocity, the appropriate orifice area may be determined from the graph. In the previous graph the straight line predicts where the average of each population would fall, and the actual sample average from the experiment falls within the range expected due to the variation of individuals.

The interpretation of ANOVA results falls into two categories initially,

1. Factors which have an F ratio exceeding some criterion
2. Factors which have an F ratio less than some criterion

The factors which have an F ratio larger than the criterion (F ratio from the tables) are believed to influence the average value for the population, and factors which have an F ratio less than the criterion are believed to have no effect on the average. A complete form of the ANOVA table would now appear as in Table 2-10. Method 2 and the format for the ANOVA summary shown in Table 2-10 will be the basic analytical technique used throughout the remainder of this text.

2-4 Two-Way ANOVA

Two-way ANOVA is the next highest order of ANOVA to review; there are two controlled parameters in this experimental situation. Method 2 will be used, but the graphical representations will be discontinued although utilization is still possible.

2-4-1 Casting experiment analysis

During one seminar a student proposed this experiment. He worked at an aluminum casting foundry which manufactured pistons for reciprocating engines. One problem existed at the end of the casting process, which was how to attain the proper hardness of the casting for a particular product. The hardness was measured on the Rockwell B scale. Engineers were interested in the effect of copper and magnesium content on casting hardness. According to specifications the copper content could be 3.5 to 4.5% and the magnesium content could be 1.2 to 1.8%. An experiment could be run to evaluate these factors and conditions simultaneously, using this symbology

$$A = \%\text{ copper content} \qquad A_1 = 3.5 \qquad A_2 = 4.5$$

$$B = \%\text{ magnesium content} \qquad B_1 = 1.2 \qquad B_2 = 1.8$$

TABLE 2-10 Complete One-Way ANOVA Summary

Source	SS	ν	V	F
A	2.0150	2	1.008	19.4#
e	.4675	9	.052	
T	2.4825	11		

[†]At least 90% confidence.
[‡]At least 95% confidence.
[#]At least 99% confidence.

TABLE 2-11   Two-Way Experimental Layout

	A_1	A_2
B_1		
B_2		

The experimental conditions can all be shown in Table 2-11. Four possible combinations exist: A_1B_1, A_1B_2, A_2B_1, and A_2B_2. Imagine that the four different mixes of metal constituents are prepared, castings poured, and resulting hardness measured. Two parts are randomly selected from each batch of castings and hardness measured. The results could quite possibly look like those in Table 2-12.

Remembering that the variation due to the mean will not be considered anyway, 70 points of hardness is subtracted from each value to simplify the discussion. Transformed results are shown in Table 2-13.

2-4-2   Sums of squares (two-way ANOVA)

In two-way ANOVA total variation may be decomposed into more components:

1. Variation due to factor A
2. Variation due to factor B
3. Variation due to the interaction of factors A and B
4. Variation due to error

An equation for total variation may be written

$$SS_T = SS_A + SS_B + SS_{A \times B} + SS_e$$

$A \times B$ represents the interaction of factors A and B. The interaction is the mutual effect of copper and magnesium in affecting casting hardness. If the effect on hardness of the percentage of copper depends on the percentage of magnesium, then an interaction is said to be present. This will be discussed in more detail in Sec. 2-4-4.

Some preliminary calculations will speed the ANOVA for this experiment, as summarized in Table 2-14.

TABLE 2-12   Two-Way Experimental Data

	A_1	A_2
B_1	76, 78	73, 74
B_2	77, 78	79, 80

TABLE 2-13   Transformed Data for Two-Way ANOVA

	A_1	A_2
B_1	6, 8	3, 4
B_2	7, 8	9, 10

$$A_1 = 29 \qquad B_1 = 21 \qquad T = 55$$

$$A_2 = 26 \qquad B_2 = 34$$

$$n_{A_1} = 4 \qquad n_{B_1} = 4 \qquad N = 8$$

$$n_{A_2} = 4 \qquad n_{B_2} = 4$$

The total variation is

$$SS_T = \left[\sum_{i=1}^{N} y_i^2\right] - \frac{T^2}{N}$$

$$= 6^2 + 8^2 + 3^2 + \cdots + 10^2 - \frac{55^2}{8} = 40.875$$

The variation due to factor A can be calculated a couple of ways. The general formula for any number of levels of factor A is

$$SS_A = \left[\sum_{i=1}^{k_A}\left(\frac{A_i^2}{n_{A_i}}\right)\right] - \frac{T^2}{N}$$

$$= \frac{A_1^2}{n_{A_1}} + \frac{A_2^2}{n_{A_2}} + \cdots + \frac{A_k^2}{n_{A_k}} - \frac{T^2}{N}$$

$$= \frac{A_1^2}{n_{A_1}} + \frac{A_2^2}{n_{A_2}} - \frac{T^2}{N}$$

TABLE 2-14   Two-Way Layout, Summarized Data

	A_1	A_1	Total	
B_1	6, 8	3, 4	21	
B_2	7, 8	9, 10	34	
Total	29	26	55	Grand total

$$= \frac{29^2}{4} + \frac{26^2}{4} - \frac{55^2}{8} = 1.125$$

This is an opportune time to point out a mathematical check of the SS_A calculation. Note that

$$29 + 26 = 55 \quad \text{and} \quad 4 + 4 = 8$$

The sum of the numerators of the plus terms (neglecting the square) must equal the numerator of the negative term. The sum of the denominators of the plus terms must equal the denominator of the negative term. If these conditions are not met, then the SS_A calculation will be wrong. For a two-level experiment when sample sizes are equal, this equation can be simplified to this special formula:

$$SS_A = \frac{(A_1 - A_2)^2}{N}$$

$$= \frac{(29 - 26)^2}{8} = \frac{3^2}{8} = 1.125$$

Similarly, the variation due to factor B is

$$SS_B = \frac{(B_1 - B_2)^2}{N}$$

$$= \frac{(21 - 34)^2}{8} = 21.125$$

To calculate the variation due to the interaction of factors A and B, the same concept of squaring the magnitude of the line segments due to the factor A and B combinations is applied but will not be graphically shown. To simply show the calculation method will suffice at this point. When the variation due to factor A was calculated, the data was organized in the fashion of Table 2-15. When variation due to factor B was calculated, the data was organized as in Table 2-16. To calculate the variation due to the interaction of factors A and B, the data must be organized in the manner shown in Table 2-17. This data is organized into all the possible factor A and B combinations and the data summed for each combination. Discussed below are two methods for calculating the

TABLE 2-15   Data Summarized over Factor A

	A_1	A_2	
4 data points make up this total	29	26	4 data points make up this total

TABLE 2-16   Data Summarized over Factor B

B_1	21	4 data points make up each
B_2	34	of these totals

variation due to an interaction. The first is a general method for any number of levels of either factor and the second is a specific method for only two level factors.

General formula.   Let $(A \times B)_i$ represent the sum of data under the ith condition of the combinations of factor A and B. Also, let c represent the number of possible combinations of the interacting factors and $n_{(A \times B)_i}$ the number of data points under this condition. Then

$$SS_{A \times B} = \left[\sum_{i=1}^{c} \left(\frac{(A \times B)_i^2}{n_{A \times B_i}} \right) \right] - \frac{T^2}{N} - SS_A - SS_B$$

So for the example problem,

$$A_1 B_2 = (A \times B)_1 = 14$$

$$A_2 B_1 = (A \times B)_2 = 7$$

$$A_1 B_2 = (A \times B)_3 = 15$$

$$A_2 B_2 = (A \times B)_4 = 19$$

$$c = 4$$

$$SS_{A \times B} = \frac{14^2}{2} + \frac{7^2}{2} + \frac{15^2}{2} + \frac{19^2}{2} - \frac{55^2}{8} - 1.125 - 21.125$$

$$= 15.125$$

Note that when the various combinations are summed, squared, and divided by the number of data points for that combination, the subsequent value also includes the factor main effects which must be subtracted. When interaction effects are calculated using the general formula, all lower-order interactions and factor effects must be subtracted.

TABLE 2-17   Data Summarized over Interaction $A \times B$

	A_1	A_2	
B_1	14	7	2 data points make up
B_2	15	19	each of these totals

Specific formula. In the test data summations were made in two directions to assess the main effects of factors A and B, vertically to obtain the effect of factor A, and horizontally to obtain the effect of factor B. In a two-factor, two-level experiment, one other summation possibility exists which assesses the interaction effect, that is, diagonally. $A_1B_1 + A_2B_2 = A{\times}B_1$ which represents one diagonal sum and $A_1B_2 + A_2B_1 = A{\times}B_2$ which represents the other diagonal sum. Since one comparison of two groups is made, the specific formula may be applied

$$\text{SS}_{A{\times}B} = \frac{(A{\times}B_1 - A{\times}B_2)^2}{N}$$

$$= \frac{(22 - 33)^2}{8} = 15.125$$

Note that this method includes the interactive effect only; the lower-order interactions and factor effects need not be subtracted.

The easiest way to calculate error variation in this case and also more complex ANOVAs is to use the lazy technique. Whatever variation is left over from that which is accountable must be error, where

$$\text{SS}_T = \text{SS}_A + \text{SS}_B + \text{SS}_{A{\times}B} + \text{SS}_e$$

Then

$$\text{SS}_e = \text{SS}_T - \text{SS}_A - \text{SS}_B - \text{SS}_{A{\times}B}$$

$$= 40.875 - 1.125 - 21.125 - 15.125 = 3.500$$

2-4-3 Degrees of freedom (two-way ANOVA). The degrees of freedom for all items except the interaction are identical to the Method 2 calculations.

Total degrees of freedom

$$\nu_T = N - 1 = 8 - 1 = 7$$

also

$$\nu_T = \nu_A + \nu_B + \nu_{A{\times}B} + \nu_e$$

Factor A degrees of freedom

$$\nu_A = k_A - 1 = 2 - 1 = 1$$

Factor B degrees of freedom:

$$\nu_B = k_B - 1 = 2 - 1 = 1$$

Interaction $(A \times B)$ degrees of freedom:

$$\nu_{A \times B} = (\nu_A)(\nu_B) = 1 \times 1 = 1$$

Error degrees of freedom:

$$\nu_e = \nu_T - \nu_A - \nu_B - \nu_{A \times B}$$

$$= 7 - 1 - 1 - 1 = 4$$

The 4 d.f. for error are readily apparent in this situation by referring to Table 2-13. One fair comparison between two data points within each treatment condition is available.

2-4-4 ANOVA summary table (two-way ANOVA)

The summary of the ANOVA results is shown in Table 2-18. The ANOVA results indicate that copper content by itself has no effect on the resultant casting hardness, magnesium content by itself has a substantial effect (largest SS) on hardness, and the interaction of copper and magnesium content plays a substantial part in determining hardness. A plot of the test data shows this in Figure 2-13.

In this plot there exist nonparallel lines which indicate the presence of an interaction. The factor A effect depends on the level of factor B and vice versa. If the lines were parallel, there would be no interaction (the factor A effect would be the same regardless of the level of factor B). The average of all results under A_1 and all results under A_2 are not very different and the average of all results under B_1 and all results under B_2 are different. The repeated data points at each interaction combination imply there is some error variation around the average for that combination. Note that in this case the conditions A_1B_1 and A_1B_2 do not appear to be different. Later, a method will be shown which will aid in making the decision of whether or not this is true.

TABLE 2-18 ANOVA of Casting Experiment

Source	SS	ν	V	F
A	1.125	1	1.125	1.29
B	21.125	1	21.125	24.14‡
$A \times B$	15.125	1	15.125	17.29†
e	3.500	4	.875	
T	40.875	7		

†At least 90% confidence.
‡At least 95% confidence.
#At least 99% confidence.

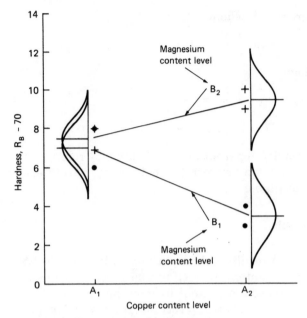

Figure 2-13 Copper-magnesium interaction.

Geometrically, there is some information available from the graph that may be useful, as seen in Fig. 2-14. The relative magnitudes of the various effects can be seen graphically. The B effect is the largest, the $A \times B$ effect next largest, and the A effect is very small.

More interpretation of this kind of information will be discussed in later chapters.

2-5 Three-Way ANOVA

One final ANOVA method will be discussed which will demonstrate some mathematical complications in ANOVA. Three-way ANOVA entails three controlled factors in an experiment.

Dispensing with any assumed experimental situation, the arrangement for the experiment would appear like Table 2-19. Here is a case where factor B has three levels and other factors two levels. The test data could be as in Table 2-20.

2-5-1 Sums of squares (three-way ANOVA)

An equation for total variation may be written in this case:

$$SS_T = SS_A + SS_B + SS_C + SS_{A \times B} + SS_{A \times C} + SS_{B \times C} + SS_{A \times B \times C} + SS_e$$

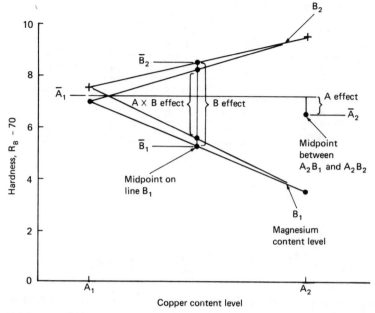

Figure 2-14 Copper-magnesium interaction; magnitude of factorial effects.

Also, the total variation may be calculated

$$SS_T = \left[\sum_{i=1}^{N} y_i^2 \right] - \frac{T^2}{N}$$

$$= 1^2 + 3^2 + 7^2 + \cdots + 15^2 + 17^2 - \frac{232^2}{24} = 457.333$$

TABLE 2-19 Three-Way Experimental Layout

	A_1	A_2	
B_1			C_1
			C_2
B_2			C_1
			C_2
B_3			C_1
			C_2

TABLE 2-20 Data for Three-Way Experiment

	A_1	A_2	
B_1	1 3	7 9	C_1
	3 5	8 12	C_2
B_2	4 8	10 12	C_1
	10 14	12 16	C_2
B_3	6 8	11 13	C_1
	13 15	15 17	C_2

To calculate the variation due to factor A, the data is first summed as shown in Table 2-21.

$$SS_A = \frac{(A_1 - A_2)^2}{N} = \frac{(90 - 142)^2}{24} = 112.667$$

The data can be organized as in Table 2-22, and the variation due to factor B determined:

$$SS_B = \frac{B_1^2}{n_{B_1}} + \frac{B_2^2}{n_{B_2}} + \frac{B_3^2}{n_{B_3}} - \frac{T^2}{N}$$

$$SS_B = \frac{48^2}{8} + \frac{86^2}{8} + \frac{98^2}{8} - \frac{232^2}{24} = 170.333$$

When the data is organized as in Table 2-23, the variation due to factor C may be calculated:

$$SS_C = \frac{(C_1 - C_2)^2}{N} = \frac{(92 - 140)^2}{24} = 96.000$$

TABLE 2-21 Factor A Effect

A_1	A_2
90	142

TABLE 2-22 Factor B Effect

B_1	48
B_2	86
B_3	98

TABLE 2-23 Factor C Effect

92	C_1
140	C_2

To calculate the interaction the data must be arranged into all the possible combinations of factors A and B and summed within those combinations as shown in Table 2-24.

$$SS_{A \times B} = \left[\sum_{i=1}^{c} \left(\frac{(A \times B)_i^2}{n_{A \times B_i}} \right) \right] - \frac{T^2}{N} - SS_A - SS_B$$

$$SS_{A \times B} = \frac{12^2 + 36^2 + 36^2 + 50^2 + 42^2 + 56^2}{4} - \frac{232^2}{24} - 112.667 - 170.333$$

$$SS_{A \times B} = 8.333$$

The variation due to the interaction of factors A and C may be determined after the data is organized into combinations of factor A and factor C shown in Table 2-25.

$$SS_{A \times C} = \frac{(122 - 110)^2}{24} = 6.000$$

Variation due to the interaction of factors B and C is calculated similarly when the six combinations of factor B and C are arranged in Table 2-26.

TABLE 2-24 A×B Interaction Effect

	A_1	A_2
B_1	12	36
B_2	36	50
B_3	42	56

TABLE 2-25 A×C Interaction Effect

	A_1	A_2	
	30	62	C_1
	60	80	C_2
122		110	

TABLE 2-26 $B \times C$ Interaction Effect

B_1	20	C_1
	28	C_2
B_2	34	C_1
	52	C_2
B_3	38	C_1
	60	C_2

$$SS_{B \times C} = \frac{20^2 + 28^2 + 34^2 + 52^2 + 38^2 + 60^2}{4} - \frac{232^2}{24} - 170.333 - 96.000$$

$$= 13.000$$

To determine the variation due to the interaction of factors A, B, and C, the data must be organized into all the possible factor A, B, and C combinations and summed for that treatment condition like the arrangement in Table 2-27.

$$SS_{A \times B \times C} = \left[\sum_{i=1}^{c} \left(\frac{(A \times B \times C)_i^2}{n_{(A \times B \times C)_i}} \right) \right] - \frac{T^2}{N} - SS_A - SS_B - SS_C$$

$$- SS_{A \times B} - SS_{A \times C} - SS_{B \times C}$$

Note that all lower-order interactions and factor effects must be subtracted from the summation.

$$SS_{A \times B \times C} = \frac{4^2 + 16^2 + 8^2 + \cdots + 32^2}{2} - \frac{232^2}{24} - 112.667$$

$$- 170.333 - 96.000 - 8.333 - 6.000 - 13.000$$

$$= 3.000$$

Error can be determined by finding the remainder of the total variation left from the known sources:

$$SS_e = SS_T - SS_A - SS_B - SS_C - SS_{A \times B} - SS_{A \times C} - SS_{B \times C} - SS_{A \times B \times C}$$

$$= 457.333 - 112.667 - 170.333 - 96.000 - 8.333$$
$$- 6.000 - 13.000 - 3.000$$

$$= 48.000$$

TABLE 2-27 $A \times B \times C$ Interaction Effect

	A_1	A_2	
B_1	4	16	C_1
	8	20	C_2
B_2	12	22	C_1
	24	28	C_2
B_3	14	24	C_1
	28	32	C_2

2-5-2 Degrees of freedom (three-way ANOVA)

The degrees of freedom for the various factors and interactions are

$$\nu_T = \nu_A + \nu_B + \nu_C + \nu_{A \times B} + \nu_{A \times C} + \nu_{B \times C} + \nu_{A \times B \times C}$$

$$= N - 1 = 24 - 1 = 23$$

$$\nu_A = k_A - 1 = 2 - 1 = 1$$

$$\nu_B = k_B - 1 = 3 - 1 = 2$$

$$\nu_C = k_C - 1 = 2 - 1 = 1$$

$$\nu_{A \times B} = (\nu_A)(\nu_B) = (1)(2) = 2$$

$$\nu_{A \times C} = (\nu_A)(\nu_C) = (1)(1) = 1$$

$$\nu_{B \times C} = (\nu_B)(\nu_C) = (1)(2) = 2$$

$$\nu_{A \times B \times C} = (\nu_A)(\nu_B)(\nu_C) = (1)(2)(1) = 2$$

$$\nu_e = \nu_T - \nu_A - \nu_B - \nu_C - \nu_{A \times B} - \nu_{A \times C} - \nu_{B \times C} - \nu_{A \times B \times C}$$

$$\nu_e = 23 - 1 - 2 - 1 - 2 - 1 - 2 - 2 = 12$$

2-5-3 ANOVA summary table (three-way ANOVA)

The ANOVA summary is shown in Table 2-28. A procedure to utilize information associated with insignificant factors and/or interactions will be discussed in later chapters.

At this point, main effects A, B, and C all appear to be significant in

TABLE 2-28   Three-way ANOVA Summary

Source	SS	ν	V	F
A	112.667	1	112.667	28.17#
B	170.333	2	85.167	21.29#
C	96.000	1	96.000	24.00#
$A \times B$	8.333	2	4.167	1.04
$A \times C$	6.000	1	6.000	1.50
$B \times C$	13.000	2	6.500	1.63
$A \times B \times C$	3.000	2	1.500	.38
e	48.000	4	4.000	
T	457.333	23		

†At least 90% confidence.
‡At least 95% confidence.
#At least 99% confidence.

affecting the average in this experiment. Plots should be made and interpreted accordingly.

2-6   Critique of the F Test

There are limitations in this method, of which the reader should be aware. Referencing the practice problem in Sec. 2-3-9, the error variance of .052 was obtained by averaging the sample variances from each of the three controlled groups of the levels of factor A. This happens automatically in ANOVA because the sums of squares for all estimates of error are pooled together and divided by the total degrees of freedom for error. A basic assumption of ANOVA is that error variance is equal for all treatment conditions (combinations of the various levels of the various factors); however, this may not be true. Because of the automatic averaging, an opportunity to reduce variation by controlling levels of design parameters may go unrecognized. Because of the loss function, opportunities to reduce variation should be sought out and utilized. Therefore, the F test will be used as a reference decision making tool and other methods will be utilized to detect a reduction in variation.

Another limitation of the F test is that only the alpha risk is addressed, which is the risk of saying a factor affects the average performance when in fact it does not. Another risk which is the opposite side of the same coin is the beta risk. The beta risk is the risk of saying a factor has no effect on the average performance when in fact it does. Since the beta risk is not assessed, the experimenter may not know what chance there is of missing an important factor. The alpha and beta risks will be discussed more in Sec. 3-4-6.

2-7 Summary

This chapter started with the very simplest ANOVA that can be done and progressed sequentially through more complex ANOVAs to three-way ANOVA. This analysis method can be applied to more complex experimental situations or to any set of observations that are structured (have controlled groups with identical operating conditions) to allow decomposition of variation into accountable sources. The ANOVA method will be applied to some very specialized situations in the next chapter on orthogonal arrays.

Problems

2-1 Perform no-way ANOVA on these data points:

Data (y): 10, 7, 9, 12, 11, 8, 9

2-2 Perform one-way ANOVA (method 2) for four types of cardboard; the data is the burst load.

A_1 12, 15, 14
A_2 19, 20, 18, 21, 22
A_3 11, 11, 12, 12
A_4 23, 24

2-3 Perform two-way ANOVA on the data arranged for factors A and B.

A_1B_1 15, 13
A_1B_2 15, 14
A_2B_1 11, 12
A_2B_2 17, 18

3

Introduction to Orthogonal Arrays

3-1 Typical Test Strategies

Engineers and scientists are most often faced with two product (or process) development situations. The terms product and process can be used interchangeably in the following discussion, because the same approaches would apply if developing either a product or process. One development situation is to find a parameter that will improve some performance characteristic to an acceptable or optimum value. A second situation is to find a less expensive, alternative design, material, or method which will provide equivalent performance. Depending on which situation the experimenter is facing, different strategies may be used. The first problem of needing to improve performance is the most typical situation.

When searching for improved or equivalent designs, the person typically runs some test, observes some performance of the product, and makes a decision to use the new design or to reject the new design. It is the quality of this decision that can be improved upon when proper test strategies are utilized, in other words, to avoid the mistake of using an inferior design or not using an acceptable design.

Before the discussion of orthogonal arrays (OAs), it would be best to

TABLE 3-1 One-Factor Experiment

Trial no.	Factor level	Test results
1	1	* *
2	2	* *

review some often used test strategies. Not being aware of efficient, proper test strategies, experimenters resort to the following approaches. The most common test plan is to evaluate the effect of one parameter on product performance. A typical progression of this approach, when the first parameter chosen doesn't work, is to evaluate the effect of several parameters on product performance one at a time. The most urgent and desperate of situations finds the experimenter usually evaluating the effect of several parameters on performance all at the same time.

These different test strategies can be symbolized in the following manner. The simplest case of testing the effect of one parameter on performance would be to run a test at two different conditions of that parameter, for example, the effect of cutting speed on the microfinish of a machined part. Two different cutting speeds could be used and the resultant microfinish measured to determine which cutting speed gave the most satisfactory results. If the first level, the first cutting speed, is symbolized by a 1 and the second level, the second cutting speed, is symbolized by a 2, the experimental conditions would appear as in Table 3-1.

The * symbolizes the values for the microfinish that would be obtained on the different test samples. The sample of 2, for example only, under trial 1 could be averaged and compared to the average of the sample of 2 under trial 2 to estimate the effect of cutting speed. To do this in a statistically proper fashion, a valid number of samples under trial 1 and trial 2 would have to be made; two under each condition may not be adequate. Most engineers are not familiar with the statistical methods of determining the proper sample sizes, but there are statistics books that can aid in that determination.*,†

If the first factor chosen fails to produce the hoped for results, the person usually resorts to testing some other factor, and the resultant test program would appear as below. In generic terms, let's assume the experimenter has looked at four different factors labeled A, B, C, and D,

---

*W. J. Diamond, *Practical Experiment Designs for Engineers and Scientists*. Lifetime Learning Publications, Belmont, Calif., 1981.

†C. Lipson, N. J. Sheth, *Statistical Design and Analysis of Engineering Experiments*. McGraw-Hill, New York.

TABLE 3-2 Several Factors One at a Time

Trial no.	A	B	C	D	Test results
		Factor and factor level			
1	1	1	1	1	* *
2	2	1	1	1	* *
3	1	2	1	1	* *
4	1	1	2	1	* *
5	1	1	1	2	* *

each evaluated one at a time. One can see in Table 3-2 that the first trial is the baseline condition. The results of trial 2 can be compared to trial 1 to estimate the effect of factor A on product performance. The results of trial 3 may be compared to trial 1 to estimate the effect of factor B on product performance, and so on. Each factor level is changed one at a time, holding all others constant. This is the traditional "scientific" approach to experimentation often taught in today's high school and college chemistry and physics classes.

The third and most urgent situation finds the person grasping at straws and changing several things all at the same time in hopes that at least one of the changes will improve the situation sufficiently. Again, one can see in Table 3-3 that the first trial represents the baseline condition. The average of the data under trial one may be compared to the average of the data under trial two to determine the combined effect of all factors.

All of the methods have some type of limitation(s). These will be discussed separately.

3-1-1 One-factor experiment

The one-factor experiment evaluates the effect of one parameter on performance while ostensibly holding everything else constant. If there happens to be an interaction of the factor studied with some other factor, then this interaction cannot possibly be observed. Also, a one-factor experiment doesn't use the data in an effective manner. If a valid sample size for each level were 4, the ANOVA for such an experiment would

TABLE 3-3 Several Factors All at the Same Time

Trial no.	A	B	C	D	Test results
		Factor and factor level			
1	1	1	1	1	* *
2	2	2	2	2	* *

TABLE 3-4 Partial ANOVA
Summary Table

Factor	SS	ν
A	XX	1
e	XX	6
T	XX	7

appear as in Table 3-4. Here, 1 degree of freedom is associated with factor A and 6 degrees of freedom are associated with error (unknown factors). In a statistical sense, the more degrees of freedom associated with an item, the more information that is known about this item's effect. In this example, something is known about the effect of factor A and a lot more is known about error. But what problems are solved by knowing a lot about error? It would be to the experimenter's advantage to trade some of the degrees of freedom, i.e., to trade error degrees of freedom for degrees of freedom for more factors while not increasing the total number of tests. It appears that with the correct test strategy, a possibility exists of evaluating seven different factors with only eight tests by trading all of the error degrees of freedom.

3-1-2 Several factors one at a time

The main limitation of several factors one at a time is that no interaction among the factors studied can be observed. Also, this strategy makes limited use of the test data when evaluating factor effects. Of the ten data points represented, only two are used to compare against two others; the remaining six data points are temporarily ignored. If an attempt is made to use all the data points, then the experiment will not be orthogonal. Orthogonality means that factors can be evaluated independently of one another; the effect of one factor does not bother the estimation of the effect of another factor. One provision of orthogonality is a balanced experiment; an equal number of samples under the various treatment conditions (an equal number of tests under A_1 and A_2).

For instance, the nonorthogonality of the data set in Table 3-2 can be seen. If all of the data under level A_1 is averaged and all the data under level A_2 is averaged and compared, this is not a fair comparison of A_1 to A_2. Of the four trials under level A_1, three were at level B_1 and one at level B_2. The one trial under level A_2 was at level B_1. Therefore, one can see that if factor B has an effect on performance it will be part of the observed effect of factor A, and vice versa. Only when trial 1 is compared to other trials one at time are the factor effects orthogonal.

3-1-3 Several factors all at the same time

This situation makes separation of any of the main factor effects impossible, let alone any interaction effects. Some factors may be making a positive contribution and others a negative contribution, but no hint of this fact will exist. How can poor utilization of test data and a non-orthogonal situation be avoided? How can the interactions be estimated and still have an orthogonal experiment? The use of full factorial experiments is one possibility. The use of some strategic orthogonal arrays is another.

3-2 Better Test Strategies

Referring to the two-way ANOVA example in Sec. 2-4-1, a full factorial experiment can be symbolized in Table 3-5. Here one can see that the full factorial experiment is orthogonal in this case. There is an equal number of test data points under each level of each factor. Note that under level A_1, factor B has two data points under B_1 condition and two under B_2 condition. The same is true under the level A_2. The same balanced situation is true when looking at the experiment with respect to the two conditions of B_1 and B_2. Because of this balanced arrangement, factor A does not influence the estimate of the effect of factor B and vice versa. In matrix algebra, there are some mathematical relationships that are true for an experiment that is orthogonal, but the most practical view of this is to observe the balanced treatments within the experiment.

One can see that all possible combinations of the two factors and the two levels are represented in the above test matrix. Using this information, both factor and interaction effects can be estimated. A full factorial experiment is acceptable when only a few factors are to be investigated, but not very acceptable when there are many factors. If a full factorial experiment is used, there is a minimum of 2^f possible combinations that must be tested (f = the number of factors each at two levels). Frequently, a typical engineering investigation may initially involve five or more factors.

An actual example of an experiment used in an engine plant to

TABLE 3-5 Full Factorial Experiment

Trial no.	A	B	y data (R_B)	
1	1	1	76	78
2	1	2	77	78
3	2	1	73	74
4	2	2	79	80

Factor and factor level

TABLE 3-6 Water Pump Leak Factors and Levels

	Factor	Level 1	Level 2
A	Front cover design	Production	New
B	Gasket design	Production	New
C	Front bolt torque	Low	High
D	Gasket coating	No	Yes
E	Pump housing finish	Rough	Smooth
F	Rear bolt torque	Low	High
G	Torque pattern	Front-rear	Rear-front

investigate the problem of water pump leaks involved seven factors. Water pump leak factors and levels are shown in Table 3-6.

If a full factorial experiment is to be used in this situation, then a total of 128 tests must be conducted. This type of experiment, shown in Fig. 3-1, estimates all the main factor effects and all the possible interactions, all orthogonal to one another. However, usual time and financial limitations preclude the use of a full factorial experiment. How can an engineer efficiently (economically) investigate these design factors?

3-3 Efficient Test Strategies

Statisticians have developed more efficient test plans, which are referred to as fractional factorial experiments (FFEs). FFEs use only a portion of the total possible combinations to estimate the main factor effects and some, not all, of the interactions. Shown in Fig. 3-2 are a one-half FFE, a one-quarter FFE, and a one-eighth FFE. Certain treatment conditions are chosen to maintain the orthogonality among the various factors and interactions. It is obvious that the one-eighth FFE with only

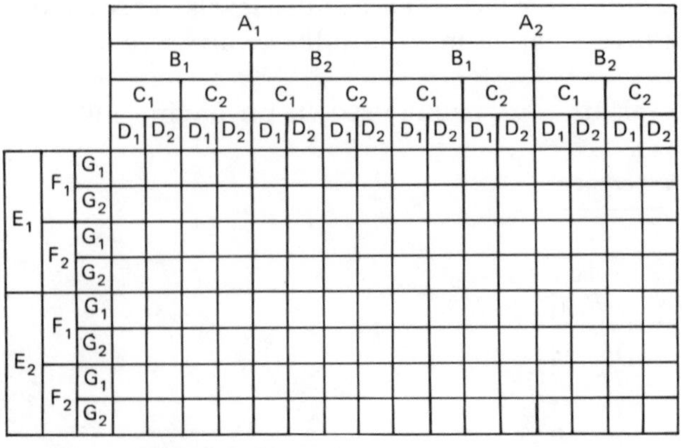

Figure 3-1 Full factorial experiment.

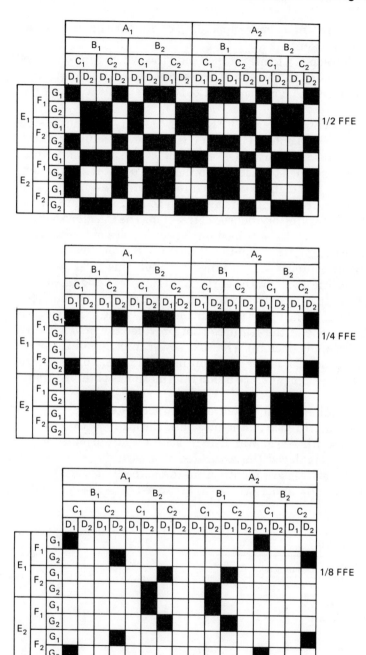

Figure 3-2 Fractional factorial experiments.

16 test combinations is much more appealing to the experimenter from a time and cost standpoint.

Taguchi has developed a family of FFE matrices which can be utilized in various situations. In this situation, a possible matrix is an eight-trial OA, which is labeled an L8 matrix. An L8, two-level matrix is shown in Table 3-7.

Actually, this is a one-sixteenth FFE which has only 8 of the possible 128 combinations represented. One can observe seven columns in this array, similar to an earlier array (several factors one at a time), which may have a factor assigned to each column. At this point, it should come as no surprise that the eight trials provide a total of seven degrees of freedom for the entire experiment, allocated to seven columns of two levels, each column having one degree of freedom. The OA allows all the error degrees of freedom to be traded for factor degrees of freedom and provides the particular test combinations that accommodate that approach. When all columns are assigned a factor, this is known as a saturated design. The levels for the particular trials are designated by 1's and 2's as before. It is easy to see that all columns provide four tests under the first level of the factor and four tests under the second level of the factor. This is one of the features that provides the orthogonality among all the columns (factors). The real power in using an OA is the ability to evaluate several factors in a minimum of tests. This is considered an efficient experiment since much information is obtained from a few trials.

OAs were a mathematical invention recorded as early as 1897 by Jacques Hadamard, a French mathematician. The utility of these arrays was not explored until World War II by Plackett and Burman, British statisticians, who employed the saturated approach described above. The Hadamard matrices are identical mathematically to the Taguchi matrices; the columns and rows are rearranged. The assignment of

TABLE 3-7   L8 OA Matrix

Trial no.	Column no.						
	1	2	3	4	5	6	7
1	1	1	1	1	1	1	1
2	1	1	1	2	2	2	2
3	1	2	2	1	1	2	2
4	1	2	2	2	2	1	1
5	2	1	2	1	2	1	2
6	2	1	2	2	1	2	1
7	2	2	1	1	2	2	1
8	2	2	1	2	1	1	2

factors to a saturated FFE is not difficult; all columns are assigned a factor. However, experiments which are not fully saturated may be more complicated to design.

3-4   Steps in Designing, Conducting, and Analyzing an Experiment

The major initial steps are

1. Selection of factors and/or interactions to be evaluated
2. Selection of number of levels for the factors
3. Selection of the appropriate OA
4. Assignment of factors and/or interactions to columns
5. Conduct tests
6. Analyze results
7. Confirmation experiment

The following sections will discuss the relevancy of these steps in performing a good experiment. Steps 1 through 4 concern the actual design of the experiment.

3-4-1   Selection of factors and/or interactions to evaluate

The determination of which factors to investigate hinges upon the product or process performance characteristic(s) or response(s) of interest. The customer who eventually uses a product expects or needs some function from a product. If during initial development stages of a product the function is not provided or consistently provided, the performance characteristic will have to be improved. Several methods are useful for determining which factors to include in initial experiments. These are

1. Brainstorming
2. Flowcharting (especially for processes)
3. Cause-effect diagrams

Brainstorming.   This activity involves bringing together a group of people associated with the particular problems and soliciting their advice concerning what to investigate. Here it is very appropriate to bring in product or process experts and statistically oriented people to discuss the factors and the structure of the experiment.

Flowcharting.   In the case of a process, flowcharts are particularly useful in the determination of factors affecting the process results. The flowchart adds some structure to the thought process and thus may avoid the omission of important factors.

An example of a process flowchart for a casting problem, to be discussed in depth in Chap. 7, is very simple. The casting process involved the steps shown in Fig. 3-3. The factors suggested from the flowcharts are

Pour casting:	Temperature of metal
	Speed of pouring
	Chemistry of metal
Cool in mold:	Time in mold
	Ambient temperature of mold
Shake-out casting:	Intensity of vibration
	Time of vibration
Air cool:	Ambient temperature
	Rate of air flow
Shot blast:	Intensity of shot blast
	Time of shot blast

The problem observed at the end of the casting process was that a percentage of the castings were cracked at various locations. The factors to include in an experiment should be the ones thought relevant to the problem of casting cracks. All factors that are thought to influence a performance characteristic should be included in the initial round of experimentation. It is better to have many factors at few levels for early experiments. The purpose of the first stages of experimentation is to eliminate many factors from contention and find those important few factors that do contribute to a product problem or contribute to product quality improvement.

Cause-effect diagram.   The structure for a cause-effect (C-E) diagram begins with the basic effect that is produced and progresses to what causes there may be for this effect. Primary, secondary, and perhaps tertiary causes are branched off the main trunk of the "effect" tree. The C-E diagramn for the casting problem would appear as in Fig. 3-4.

Again, the selection of factors to be included in the experiment should

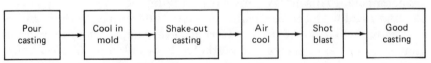

Figure 3-3   Casting process flowchart.

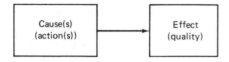

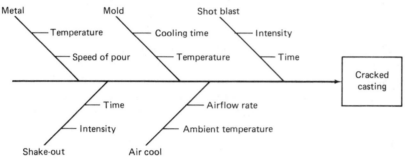

Figure 3-4 Casting process cause-effect diagram.

depend on which might affect casting cracks. Ishikawa* provides several suggestions for developing C-E diagrams. After factors are selected, any interactions that are of interest should be noted.

Not a lot of emphasis will be placed on factor selection, because most engineers and scientists have specialized knowledge concerning their particular product and have, therefore, the best background for the selection process. What is needed is the proper structure of an experiment for the factors chosen. Experimental design is a universal knowledge that can be applied to a wide range of products and processes.

3-4-2 Selection of number of levels

Initial rounds of experimentation should involve many factors at few levels; two are recommended to minimize the size of the beginning experiment. Recall that the number of degrees of freedom for a factor is the number of levels minus one; increasing the number of levels for a factor increases the total degrees of freedom in the experiment, which is a direct function of the total number of tests. One degree of freedom for each factor minimizes the total number of tests. The initial round of experimentation will eliminate many factors from contention and the few remaining can then be investigated with multiple levels without causing an undue inflation in the size of the experiment, which increases cost and/or time.

---

*K. Ishikawa, *Guide to Quality Control*. Asian Productivity Organization, Tokyo, 1976, chap. 3.

Two kinds of parameters exist which may influence a product response, continuous and discrete parameters. Continuous parameters may be measured on a scale from a very low value to a very high value and may assume any value in between, depending on the precision of the measuring instrument. Examples are temperature, speed, pressure, and time. Discrete parameters may only assume particular values such as off or on, material A, B, or C, or engine cylinder number 1, 2, 3, or 4. If continuous parameters are being used, then the initial experiment should be at two levels only; interpolation or extrapolation may be used to predict other levels. If discrete factors are used, then interpolation or extrapolation may be meaningless. For instance, the use of three different materials is possible; there is no way to interpolate or extrapolate in order to predict results of a fourth possible material. If discrete parameters are studied, then more than two levels may be required in initial experiments.

3-4-3   Selection of the OA

Degrees of freedom.   The selection of which OA to use depends on these items:

1. The number of factors and interactions of interest
2. The number of levels for the factors of interest

These two items determine the total degrees of freedom required for the entire experiment. Recalling the ANOVA chapter, the degrees of freedom for each factor is the number of levels minus one.

$$\nu_A = k_A - 1$$

The degrees of freedom for an interaction is the product of the interacting factor's degrees of freedom.

$$\nu_{A \times B} = (\nu_A)(\nu_B)$$

The minimum required degrees of freedom in the experiment is the sum of all the factor and interaction degrees of freedom.

Orthogonal arrays.   Two basic kinds of OAs are listed in the appendixes. Appendix B contains two-level arrays:

$$L4 \quad L8 \quad L12 \quad L16 \quad L32$$

Appendix C contains three-level arrays:

$$L9 \quad L18 \quad L27$$

The number in the array designation indicates the number of trials in the array; an L27 has 27 trials, for example. The total degrees of freedom available in an OA is equal to the number of trials minus one.

$$\nu_{LN} = N - 1$$

Selection of OA.   When a particular OA is selected for an experiment the following inequality must be satisfied.

$$\nu_{LN} \geq \nu_{\text{required for factors and interactions}}$$

The number of levels used in the factors should be used to select either two- or three-level kinds of OAs. If the factors are two-level, then an array from Appendix B should be chosen; if factors are three-level, then an array from Appendix C should be chosen. If some factors are two-level and some three-level, then whichever is predominant should indicate which kind of OA is selected. Chapter 4 will discuss how to modify basic OAs to accommodate a mixture of two-, three-, and four-level factors.

Once the decision is made between a two-level or three-level OA, then the number of trials for that kind of array must provide an adequate total degrees of freedom. Many times the required degrees of freedom will fall between the degrees of freedom provided by two of the OAs. The next larger OA must then be chosen.

Occasionally, the interactions to be evaluated will require an even larger OA to be used. For instance, if four main effects and two interactions, $A \times B$ and $C \times D$, are desired, then this will not fit into an L8 without causing confounding between $A \times B$ and $C \times D$. The required linear graph would appear as two parallel lines, but the L8 linear graphs are not structured in that manner. An L16 linear graph, type e, would be appropriate if the $A \times B$ and $C \times D$ interactions must be separated.

Once the appropriate OA has been selected, the factors and interactions can be assigned to the various columns.

3-4-4 Assignment of factors and interactions

Before getting into the detail of using some methods of assigning factors and interactions, a demonstration of a mathematical property of OAs is in order. OAs have several columns available for assignment of factors and some columns subsequently can estimate the effect of interactions of those factors.

Demonstration of interaction columns.   The simplest OA is an L4, which has the arrangement shown in Table 3-8. The two-factor experiment in

TABLE 3-8 L4 OA

	Column no.		
Trial no.	1	2	3
1	1	1	1
2	1	2	2
3	2	1	2
4	2	2	1

Sec. 2-4-1 (two-way ANOVA) can be adapted to the L4 OA. Factor A can be assigned to column 1 and factor B can be assigned to column 2. The first trial then represents the A_1B_1 condition, which has results of 6 and 8. Trial 2 represents the A_1B_2 condition, with results of 7 and 8. The entire experiment is shown in Table 3-9.

ANOVA of Taguchi L4 OA.   Typically, OAs are analyzed in the same manner as other structured experiments. Recalling the ANOVA in Sec. 2-4-2 for the example in Sec. 2-4-1.

$$SS_T = 40.875$$

$$SS_A = 1.125$$

$$SS_B = 21.125$$

$$SS_{A \times B} = 15.125 \text{ and}$$

$$SS_e = 3.500$$

The ANOVA for an OA is conducted by calculating the sums of squares for each column. The formula for SS_A is the same as in Sec. 2-4-2. The sums of squares for factor A, column 1, is

$$SS_A = \frac{(A_1 - A_2)^2}{N}$$

A_1 and A_2 are the sums of the data associated with the first and second levels of factor A, respectively

$$A_1 = 6 + 8 + 7 + 8 = 29$$

$$A_2 = 3 + 4 + 9 + 10 = 26$$

$$SS_A = \frac{(29 - 26)^2}{8} = 1.125$$

TABLE 3-9   L4 OA with Casting Data

	Factors				
	A	B			
	Column no.			y data	
Trial no.	1	2	3	$(R_B - 70)$	
1	1	1	1	6	8
2	1	2	2	7	8
3	2	1	2	3	4
4	2	2	1	9	10

The sums of squares for factor B, column 2, is calculated in the same manner

$$SS_B = \frac{(B_1 - B_2)^2}{N}$$

$$B_1 = 6 + 8 + 3 + 4 = 21$$

$$B_2 = 7 + 8 + 9 + 10 = 34$$

$$SS_B = \frac{(21 - 34)^2}{8} = 21.125$$

Note, the sums of squares for factors A and B are identical to Sec. 2-4-2.
The sums of squares for column 3 is

$$SS_3 = \frac{(3_1 - 3_2)^2}{N} = \frac{(33 - 22)^2}{8} = 15.125$$

Note, this value is equal to the $SS_{A \times B}$, which is not coincidental but is a mathematical property of the OA. The calculation is a demonstration, not a proof, that the third column represents the interaction of the factors assigned to the first and second columns. How to determine interaction columns will be covered in this chapter in the following section.

This particular L4 example is similar to the two-way ANOVA example, since each trial has two test results which provide an estimate of error variance with 4 degrees of freedom. The sums of squares due to error can be calculated:

$$SS_e = SS_T - SS_A - SS_B - SS_{A \times B}$$

$$= 40.875 - 1.125 - 21.125 - 15.125 = 3.500$$

The L4 OA having two factors assigned to it is equivalent to a full factorial experiment and the ANOVA is equivalent to a two-way ANOVA because certain columns in OAs represent the interaction of two other columns.

Location of interaction columns.   Taguchi has provided two tools to aid in the assignment of factors and interactions to arrays

1. Linear graphs
2. Triangular tables

Each OA has a particular set of linear graphs and a triangular table associated with it. The linear graphs indicate various columns to which factors may be assigned and the columns subsequently evaluate the interaction of those factors. The L4 example in Sec. 3-4-4 ANOVA demonstrated that the interaction of a factor assigned to column 1 and a factor assigned to column 2 is evaluated in column 3. The triangular tables contain all the possible interactions between factors (columns). Also, listed in Appendixes B and C are the linear graphs and triangular tables for the two-level and three-level arrays, respectively.

Linear graphs (two-level OAs).   The simplest OA, an L4, has a linear graph that appears in Fig. 3-5. Recall, the L4 OA has four trials and three columns. The linear graph indicates that factor A may be assigned to column 1, factor B to column 2, and the $A \times B$ interaction subsequently located in column 3. The dot represents a column available for a two-level factor which is allocated 1 degree of freedom. The line represents a column which will evaluate the interaction of the factors assigned to the respective dots. The interaction is also allocated 1 degree of freedom in this case.

The next most complicated linear graphs are for an L8 OA. There are two linear graphs available for an L8 as shown in Fig. 3-6. These two linear graphs indicate that several factors may be assigned to different columns and several different interactions may be evaluated in different columns. For instance, in the type b linear graph, factors A, B, C, and D may be assigned to columns 1, 2, 4, and 7, respectively. This places the $A \times B$ interaction in column 3, the $A \times C$ interaction in column 5, and the $A \times D$ interaction in column 6. The other linear graph provides an alternative arrangement with other interactions assigned.

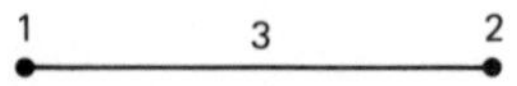

Figure 3-5 L4 linear graph.

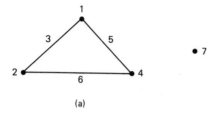

(a)

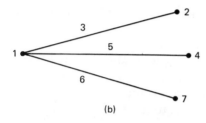

(b)

Figure 3-6 L8 linear graphs.

Triangular tables (two-level OAs). The triangular tables list all of the possible interacting column relationships that exist for a given OA. An L4 triangular table is shown in Table 3-10. The first factor assigned to an OA may actually be placed in any column, column 2 as an example. The second factor may be assigned to any other column, column 3 as an example. If factor A is assigned to column 2 and factor B assigned to column 3, the triangular table indicates the $A \times B$ interaction will be in column 1 as shown in Table 3-11. The triangular table shows the three columns are mutually interactive; 1 and 2 interact in 3, 2 and 3 interact in 1, and 1 and 3 interact in 2. Any assignment of factors A and B is mathematically and statistically equivalent. All of the two-level OA linear graphs and triangular tables function in this same manner. Note, the smaller OA linear graphs and triangular tables are a portion of the larger OA linear graphs and triangular tables.

Linear graphs (three-level OAs). An L9 OA linear graph is shown in Fig. 3-7. The dots represent columns available for a three-level factor which is allocated 2 degrees of freedom. The line represents the two columns, which together evaluate the interaction of the dot columns. Recall that

TABLE 3-10 L4 Triangular Table

Column	2	3
1	3	2
2		1

TABLE 3-11   L4 Triangular Table with Factors

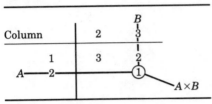

the interaction will require 4 degrees of freedom, hence the two columns are necessary.

Triangular tables (three-level OAs).   The triangular table for an L9 appears in Table 3-12. The table again indicates all the possible interacting column relationships (pairs of columns) that evaluate the interaction.

All the three-level OA linear graphs and triangular tables are used in the same manner.

Confounding of main effects and interactions.   As was discussed earlier, Plackett and Burman's use of OAs resulted in the assignment of factors to all columns. Obviously, many interactions are confounded (mixed) with the main effects. This is the major compromise of using FFEs—to reduce the number of tests some information must be surrendered.

An L8 OA provides a complex enough array to demonstrate the amount of confounding that may occur in an experiment. It is obvious that, if two factors are assigned to an L8, columns 1 and 2 are automatically available for use, but what happens as more factors are added to the experiment? If three factors are assigned (A, B, and C), the linear graph for an L8 indicates the assignment to columns 1, 2, and 4 located at the vertices of the triangle shown in Fig. 3-6a. The triangular table (see table in Appendix B) locates the $A \times B \times C$ three-factor interaction by finding the interaction of factor A and $B \times C$ interaction in columns 1 and 6, respectively. Other combinations indicate the same:

Factor and interaction columns
$$A \times (B \times C) = A \times B \times C$$
$$1 \qquad 6 \qquad\qquad 7$$

TABLE 3-12   L9 Triangular Table

Column	2	3	4
1	3,4	2,4	2,3
2		1,4	1,3
3			1,2

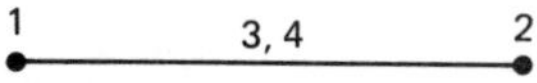

Figure 3-7 L9 linear graph.

Factor and interaction columns
$$B \times (A \times C) = A \times B \times C$$
$$2 \qquad\quad 5 \qquad\qquad 7$$

Factor and interaction columns
$$C \times (A \times B) = A \times B \times C$$
$$4 \qquad\quad 3 \qquad\qquad 7$$

The resultant column assignment for the factors and the interactions of this experiment are shown in Table 3-13.

In this situation, all main effects and all interactions can be estimated, which results in a high-resolution experiment. Resolution power indicates the clarity of which individual factors and interactions may be seen (evaluated) in an experiment. Table 1 in Appendix D indicates this experiment to be of resolution 4, which is also a full factorial experiment in this case.

If another factor, D, is added to this experiment, the most logical column assignment is 7, automatically confounding D with a three-factor interaction. Any other choice confounds D with a two-factor interaction. The three-factor interaction is much less likely to occur, and if it does, then it will most likely be of smaller magnitude than a main effect or a two-factor interaction. Now the factors and interactions appear as in Table 3-14. The resolution power of the experiment is subsequently much lower, resolution 2, because of the greater amount of confounding in the columns. Also, an interaction, $A \times B \times C \times D$, cannot be estimated in this experiment because factor D was intentionally confounded with the $A \times B \times C$ interaction.

TABLE 3-13 Resolution 4 Experiment

			Column no.			
1	2	3	4	5	6	7
A	B	$A \times B$	C	$A \times C$	$B \times C$	$A \times B \times C$

TABLE 3-14 Resolution 2 Experiment

			Column no.			
1	2	3	4	5	6	7
A	B	$A \times B$	C	$A \times C$	$B \times C$	$A \times B \times C$
$B \times C \times D$	$A \times C \times D$	$C \times D$	$A \times B \times D$	$B \times D$	$A \times D$	D

TABLE 3-15   Two-Factor Experiment

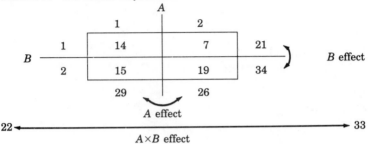

This confounding of groups of two-factor interactions can be demonstrated with the L8 OA. Recalling from example 2-4, two-way ANOVA, the factor and interaction effects are estimated in the manner shown in Table 3-15. The data is summed horizontally, vertically, and diagonally. Treatment conditions A_1B_1 and A_2B_2 represent interaction condition $A \times B_1$; A_1B_2 and A_2B_1 represent $A \times B_2$. This is seen in the L8 OA also (see Table 3-16). If the same representation is used for factors C and D, then

C_1D_1 and C_2D_2 are represented by $C \times D_1$

C_1D_2 and C_2D_1 are represented by $C \times D_2$

One can observe that the third column follows the same pattern whether using the $A \times B$ interaction or the $C \times D$ interaction, thus the confounding of the two interactions and the analytical impossibility of separating the interaction effects from each other.

If a fifth factor is added, columns 3, 5, and 6 are available, then the resolution power will drop off further to a resolution 1 because of a factor

TABLE 3-16   Resolution 2 Experiment Confounding Demonstration

			Factors and interactions				
			$A \times B$		$A \times C$	$A \times D$	
	A	B	$C \times D$	C	$B \times D$	$B \times C$	D
				Column no.			
Trial no.	1	2	3	4	5	6	7
1	1	1	1	1	1	1	1
2	1	1	1	2	2	2	2
3	1	2	2	1	1	2	2
4	1	2	2	2	2	1	1
5	2	1	2	1	2	1	2
6	2	1	2	2	1	2	1
7	2	2	1	1	2	2	1
8	2	2	1	2	1	1	2

TABLE 3-17   Resolution 1 Experiment

Column no.						
1	2	3	4	5	6	7
A	B	$A \times B$	C	$A \times C$	$B \times C$	$A \times B \times C$
$B \times C \times D$	$A \times C \times D$	$C \times D$	$A \times B \times D$	$B \times D$	$A \times D$	D
$B \times E$	$A \times E$	E	$D \times E$	$A \times D \times E$	$B \times D \times E$	$C \times E$
				$B \times C \times E$	$A \times C \times E$	

being confounded with a two-factor interaction. Since factor E was confounded with interactions $A \times B$ and $C \times D$, the three-factor interactions $A \times B \times E$ and $C \times D \times E$ cannot be evaluated. This situation is shown in Table 3-17.

Table D.2 in Appendix D lists the column assignments that should be used to provide at least a resolution 2 experiment for two-level factors. If four or fewer factors are evaluated, then columns 1, 2, 4, and 7 of an L8 should have factors assigned to them. If eight factors are evaluated, then the columns indicated under an L16 should have factors assigned to them. The key thing to remember is that as factors are added to a given OA, they should be placed in columns in which the lowest-order interaction is still of higher order than other columns. In the previous example (refer to Table 3-14), factor D was added into the three-factor interaction column of the experiment. Using this approach protects the resolution power of the experiment as factors are added.

Taguchi does not place much emphasis on the confounding that exists in a low resolution experiment; however, it is good for the experimenter to be aware of such situations. A list of factors and interactions of interest might include factors A, B, and C plus interactions $A \times B$ and $A \times C$. These can be allocated to separate columns and they appear to be completely independent from any confounding, but this is not the case. Taguchi views interactions as being unimportant because to obtain the interactive effect the experimenter must control two main effects. Since one or more main effects usually need to be controlled for a product or process anyway, the interaction causes no additional complications. In the case of an HB or LB characteristic, the main effects will indicate the proper combination of levels when the interaction effect is less than either main effect, therefore, knowledge of the interaction is irrelevant. The experimenter needs to know about the interaction only to be able to obtain the desired average response.

From a very practical point of view, all factors assigned to an experiment will not be equally influential in changing the average response. Therefore, what few factors are significant will have been evaluated as if the experiment were full factorial. The experimenter starts with what appears to be a low-resolution experiment, but when the trials are com-

Figure 3-8 Required interactions linear graph.

pleted, the practical result will be a higher-resolution experiment. If factors *A* and *B* are influential, four treatment conditions have been tested, which makes the experiment full factorial for those factors. The dilemma at the start of experimentation is not knowing which factors are really influential.

Modification of linear graphs.   Linear graphs are a tremendous help in quickly assigning factors and interactions to columns. Linear graphs can be modified to suit particular circumstances. For instance, a list of factors and interactions might include

$$A, B, C, D, E \quad C{\times}D \quad C{\times}E$$

The modification of linear graphs allows these factors and interactions to be quickly assigned to separate columns.

The first step is to draw the linear graph required for the experiment beginning with interactions. In this case the required linear graph starts with the form in Fig. 3-8. The noninteracting factors are then added as additional separate column locations, making the complete required linear graph appear as in Fig. 3-9. This linear graph shows that two factors have a possibility of an interaction with a third factor. Two other factors have no interaction of interest. If the factors are all two level, then there are a total of 7 degrees of freedom for the experiment, so an L8 may be used. One form of an L8 linear graph, type *a*, takes the form shown in Fig. 3-6.

The L8 linear graph needs to be modified to fit this particular experiment and can be modified as shown in Fig. 3-10. The 1 degree of freedom (column 6) can be "borrowed" from the one portion of the linear graph to "create" an extra dot for a noninteracting factor. Any of the interaction columns can be used, and the resultant shape of the linear graph is congruent with that required for the experiment. However, the experi-

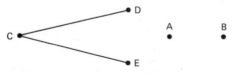

Figure 3-9 Complete required linear graph.

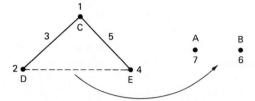

Figure 3-10 Modified L8 linear graph.

menter should realize that column 6 still contains the $D \times E$ interaction, which is obviously confounded with factor B.

Linear graphs may be modified by dismantling certain parts and reconstructing with other parts by obeying the valid interactions listed in the triangular tables. The two different appearing L8 linear graphs can demonstrate the reconstruction principle. The interaction column 6 can be used to portray an interaction between column 1 and column 7 as shown in Fig. 3-11.

Now the modified linear graph is identical to the second form of an L8 linear graph. This shows the two L8 linear graphs are really identical experiments.

Other linear graphs can be modified to suit the situation. The only procedure that must not be violated is the reconstruction according to the triangular table for that array. The triangular tables list all the possible mathematical column relationships which exist and cannot be changed.

3-4-5 Conducting the experiment

Once the factors are assigned to a particular column of the selected OA, the test strategy has been set and physical preparation for performing the test can begin. Some decisions need to be made concerning the order of testing the various trials.

Trial test conditions. Factors are assigned to columns; trial test conditions, however, are dictated by the rows. Looking again at Table 3-16, one

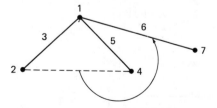

Figure 3-11 Conversion of L8 linear graph from type a to type b.

can observe that trial 6 requires the test conditions of $A_2B_1C_2D_1$. Trial 3 requires test conditions of $A_1B_2C_1D_2$. The interaction conditions cannot be controlled when conducting a test because they are dependent upon the main factor levels. Only the analysis is concerned with the interaction columns. Therefore, it is recommended that test sheets be made up which show only the main factor levels required for each trial. This will minimize mistakes in conducting the experiment which may inadvertently destroy the orthogonality.

Randomization. The order of performing the tests of the various trials should include some form of randomization. The randomized trial order protects the experimenter from any unknown and uncontrolled factors that may vary during the entire experiment and which may influence the results. Presuming that the random order does not match any of the patterns described by the columns, such as all odd-numbered trials are tested and then all even-numbered trials (matches the pattern of column 3 of an L8; all 1s and then all 2s), any unknown or uncontrolled factor effects will be spread evenly over the entire experiment. This will prevent bias of the factors and interactions assigned to the columns.

Randomization can take many forms, but the three most used approaches will be discussed.

1. Complete

2. Simple repetition

3. Complete within blocks

TABLE 3-18 **Complete Randomization (Sequentially)**

	Factors							
	A	B	C	D	E	F	G	
	Column no.							
Trial no.	1	2	3	4	5	6	7	y data
1	1	1	1	1	1	1	1	* * *
2	1	1	1	2	2	2	2	* * *
3	1	2	2	1	1	2	2	* * *
4	1	2	2	2	2	1	1	* * *
5	2	1	2	1	2	1	2	* * *
6	2	1	2	2	1	2	1	* * *
7	2	2	1	1	2	2	1	* * *
8	2	2	1	2	1	1	2	* * *

Randomized within first round———
Randomized within second round———
Randomized within third round———

Complete randomization. Complete randomization means any trial has an equal chance of being selected for the first test. To determine which trial to run next, random number tables, a random number generator, or simply drawing numbers from a hat will suffice. However, even complete randomization may have a strategy applied to it. For instance, several repetitions of each trial may be necessary, so each trial should be randomly selected until all trials have one test completed. Then each trial is randomly selected in a different order until all trials have two tests completed. The experiment will progress on a sequential basis with the opportunity for analysis at the end of each round of repetitions. This method is used when a change of test setup is very easy or inexpensive and the resulting test pattern is shown in Table 3-18.

Simple repetition. Simple repetition means that any trial has an equal opportunity of being selected for the first test, but once that trial is selected all the repetitions are tested for that trial. This method is used if test setups are very difficult or expensive to change. The simple repetition method is shown in Table 3-19.

Complete randomization within blocks. Complete randomization within blocks is used where one factor may be very difficult or expensive to change the test setup for, but others are very easy. If factor A were difficult to change, then the experiment could be completed in two halves or blocks. All A_1 trials could be randomly selected to be conducted and then all A_2 trials could be randomly selected. This arrangement is shown in Table 3-20.

Blocking known sources of variation. Randomization can be detrimental in some experimental situations. If some sources of variation are known, such as having to use two bags of plastic pellets for an injection molding experiment or two automobiles and drivers for a manual transmission

TABLE 3-19 Simple Repetition

				Factors				
	A	B	C	D	E	F	G	
Trial no.	1	2	3	4	5	6	7	y data
1	1	1	1	1	1	1	1	* * *
2	1	1	1	2	2	2	2	* * *
3	1	2	2	1	1	2	2	* * *
4	1	2	2	2	2	1	1	* * *
5	2	1	2	1	2	1	2	* * *
6	2	1	2	2	1	2	1	* * *
7	2	2	1	1	2	2	1	* * *
8	2	2	1	2	1	1	2	* * *

Random order of trials; do all repetitions for a trial

TABLE 3-20 Complete Randomization (Sequentially)

			Factors					
	A	B	C	D	E	F	G	
			Column no.					
Trial no.	1	2	3	4	5	6	7	y data
1	1	1	1	1	1	1	1	* * *
2	1	1	1	2	2	2	2	* * *
3	1	2	2	1	1	2	2	* * *
4	1	2	2	2	2	1	1	* * *
5	2	1	2	1	2	1	2	* * *
6	2	1	2	2	1	2	1	* * *
7	2	2	1	1	2	2	1	* * *
8	2	2	1	2	1	1	2	* * *

Random order of trials in block 1
Random order of trials in block 2

shiftability experiment, then these sources of variation should be treated as factors and assigned to columns. This method of controlling known sources of variation is called "blocking." If the experimenter chooses to randomize the experimental trials with respect to these known sources of variation, then the error variance will be inflated if these sources really do cause a difference in product response. If the sources are blocked and do have an effect on response, then this effect can be estimated and error variance will not be unduly inflated, which will protect the statistical power of the experiment. The rule of thumb is to block (treat as another factor) known sources of variation in an experiment. Taguchi has developed a different approach to handling this situation which will be discussed in Sec. 8-2 concerning parameter design.

Effects of randomization methods on error variance. The different methods of randomization affect error variance in different ways. Complete randomization allows a longer time to transpire between repetitions in a given trial compared to simple repetition. Because of this, unknown and uncontrolled factors that may be varying during an experiment may make the repetition-to-repetition variation larger with complete randomization compared to simple repetition. Simple repetition, because of the generally longer times transpiring between trials, will show larger trial-to-trial variation compared to complete randomization. Increased trial-to-trial variation with decreased repetition variation will tend to make factors appear significant in ANOVA when in fact they are not, so complete randomization is recommended where possible.

Selection of sample size.   From a very practical viewpoint, a minimum of one test result for each trial is required to maintain the sample size balance (orthogonality) of the experiment (if the test results are unbalanced, then the experiment requires a special analysis not covered in this text). More than one test per trial can be used, which increases the sensitivity of the experiment to detect small changes in averages of populations. An economic consideration also can be made at this time. If tests are very expensive, then one test per trial can be used. If tests are inexpensive, then more than one per trial can be used. An L8 OA with one test per trial (four tests versus four tests) makes the experimenter 90% sure (confident) of detecting a change in average of approximately two standard deviations (S). An L8 OA with two repetitions or an L16 OA with one repetition per trial (8 versus 8) makes the experimenter 90% sure of detecting a change of approximately $1\frac{1}{3}$ standard deviations. An L16 OA with two repetitions per trial provides 90% confidence in detecting a change in average of about one standard deviation. This is a fairly sensitive experiment, and sample sizes larger than this do not add much to the sensitivity. Experiments as small as an L4 OA should be avoided for the main array; the sensitivity with only one test per trial is about $3\frac{3}{4}$ standard deviations at 90% confidence. If more repetitions are added for sensitivity, then an L8 OA should be used which will evaluate interactions as well as provide the increased sensitivity with the same number of total tests.

Observed variation in an experiment.   If an experiment is conducted on a product or process where there is a history of the problem being investigated, then the variation in the experimental data should cover at least 75% of the variation seen historically. The range in the experiment from good to bad results should be at least 75% of the range from good to bad results in recent production data. The reason is to have some indication that the correct factors and levels were included in the experiment.

If the range of variation is too small, then the indication is that the factor which really causes variation has been left out of the experiment or the levels chosen for that factor in the experiment have been selected too closely. The experimenter may not be able to control or reduce variation sufficiently with the chosen factors and levels.

If the range of variation is greater than 75% of the past, then this still does not completely ensure that the truly influential factors have been included. The factor(s) which really cause variation, but are possibly included in the experiment, could have been varying unknowingly to provide the high to low values of test data. However, if it turns out the correct factors and levels were chosen, then the experimenter knows the variation can be controlled or reduced substantially.

TABLE 3-21   Two-Factor L8 OA with Data

	Factors and interaction							
	A	B	$A{\times}B$					
	Column no.							
Trial no.	1	2	3	4	5	6	7	y data (R_B-70)
1	1	1	1	1	1	1	1	6
2	1	1	1	2	2	2	2	8
3	1	2	2	1	1	2	2	7
4	1	2	2	2	2	1	1	8
5	2	1	2	1	2	1	2	3
6	2	1	2	2	1	2	1	4
7	2	2	1	1	2	2	1	9
8	2	2	1	2	1	1	2	10

3-4-6   Analysis of experimental results

This simple example is intended to demonstrate another basic property of OAs, which is that the total variation can be accounted for by summing the variation from all columns. One commonly used Taguchi OA is an L8, shown earlier in Table 3-5. This OA also can be used to display the experimental structure used in the example in Sec. 2-4-1. Factor A can be assigned to column 1 and factor B assigned to column 2 of the L8 OA. The first two trials of the OA now represent the A_1B_1 condition, which has corresponding results of 6 and 8 in the example. Trials 3 and 4 represent the A_1B_2 condition, which has results corresponding to 7 and 8. The complete OA and results are shown in Table 3-21.

ANOVA of Taguchi L8 OA.   Again, the ANOVA for an OA is conducted by calculating the sums of squares for each column. The formula for SS_A is the same as Sec. 2-4-1 and the L4 example in Sec. 3-4-4. The sums of squares for factor A, column 1, is

$$SS_A = \frac{(A_1 - A_2)^2}{N}$$

A_1 and A_2 are the sums of the data associated with the first and second levels of factor A, respectively,

$$A_1 = 6 + 8 + 7 + 8 = 29$$

$$A_2 = 3 + 4 + 9 + 10 = 26$$

$$SS_A = \frac{(29 - 26)^2}{8} = 1.125$$

The sums of squares for factor B, column 2, is calculated in the same manner

$$SS_B = \frac{(B_1 - B_2)^2}{N}$$

$$B_1 = 6 + 8 + 3 + 4 = 21$$

$$B_2 = 7 + 8 + 9 + 10 = 34$$

$$SS_B = \frac{(21 - 34)^2}{8} = 21.125$$

The sums of squares for column $A \times B$ is

$$SS_{A \times B} = \frac{(3_1 - 3_2)^2}{N} = \frac{(33 - 22)^2}{8} = 15.125$$

Note, the sums of squares for factors A and B and interaction $A \times B$ are identical to the example in Sec. 2-4-1. Continuing with the sums of squares calculations,

$$SS_4 = \frac{(4_1 - 4_2)^2}{N} = \frac{(25 - 30)^2}{8} = 3.125$$

$$SS_5 = \frac{(5_1 - 5_2)^2}{N} = \frac{(27 - 28)^2}{8} = .125$$

$$SS_6 = \frac{(6_1 - 6_2)^2}{N} = \frac{(27 - 28)^2}{8} = .125$$

$$SS_7 = \frac{(7_1 - 7_2)^2}{N} = \frac{(27 - 28)^2}{8} = .125$$

$$SS_e = SS_4 + SS_5 + SS_6 + SS_7 = 3.500$$

Note, the total of the sums of squares for the unassigned columns is equal to error sums of squares. The unassigned columns in an OA represent an estimate of error variation. Also, as in Sec. 2-4-1, there are four columns (4 degrees of freedom) associated with error variation. Here the difference of the particular array selected for the experiment changes the analysis approach slightly. The L4 has two data points per trial and the L8 one data point per trial. The error variance of the L4 must come from the repetitions in each trial, but the error variance in the L8 must come from the columns since there are no repetitions within trials.

Note also, the total of all the column sums of squares for the L8 OA is equal to SS_T for the eight data points. This is a demonstration of the property of the total sums of squares being contained within the columns of an OA.

$$SS_T = \sum SS_{columns}$$

In summary, we see the sums of squares and the degrees of freedom associated with each component of variation correlates exactly whether a 2 × 2 factorial arrangement or a two-factor, two-level OA. In this case the ANOVA for the OA is identical to a two-way ANOVA, which means that this OA example is really a full factorial experiment.

Column estimates of error variance.   In the previous example the unassigned columns were shown to estimate error variance when there was only one test result per trial. This approach of using columns to estimate error variance will be used even if all columns have factors assigned to them.

Recalling the ANOVA chapter, the variance due to a factor is really an estimate of the variance of individual data points (same as error variance) based on the variance of the sample averages of that factor. When assigned columns are used, it is obvious the variation of the average of the data associated with level 1 and level 2 can only be attributed to unknown and uncontrolled factors. In fact, the level 1s and levels 2s only represent groups of data points; when no factor or interaction is present in a column, the 1s and 2s are meaningless. Hopefully, this variation will be small; excessive variation will indicate that a potentially important factor has been excluded from the experiment.

When factors are assigned to all columns, error variance may still be estimated. Some factors assigned to an experiment will not be significant at all, even though thought to be before experimentation. This would be equivalent to saying the color of a car can affect fuel economy and assigning two different colors to a column. It is very likely this column will have a small sums of squares because it will really be estimating error variance rather than any true color effect.

When column effects turn out small in an OA then there are several possibilities:

1. No assigned factors or interactions, true error estimate

2. No significant factor and/or interaction effect

3. Very small factor and/or interaction effects

4. Canceling factor and/or interaction effects

With a low-resolution OA there is no proof that any one of the last three conditions is true or untrue, but either the second or third possibility will be accepted as true when column effects are small. From a practical point of view, there is no difference between situation 2 or 3.

A fully saturated OA will depend on some column effects turning out small relative to others and using the smaller ones as estimates of error variance.

Pooling estimates of error variance.   In the ANOVA of Sec. 3-4-4 there were four unassigned columns having 4 degrees of freedom, one for each column, which provided estimates of error variance. A better estimate was the combination of all four column effects for one overall estimate of error variance with 4 degrees of freedom. The combining of column effects to better estimate error variance is referred to as "pooling."

Two pooling strategies exist:

1. Pooling up (Taguchi)

2. Pooling down*

The pooling up strategy entails F-testing the smallest column effect against the next larger one to see if significance exists. If no significant F ratio exists, then these two effects are pooled to test the next larger column effect until some significant F ratio exists.

The pooling down strategy entails pooling all but the largest column effect and F-testing the largest against the remainder pooled together. If that column effect is significant, then the next largest is removed from the pool and those two column effects are F-tested against all others pooled until some insignificant F ratio is obtained.

The pooling up strategy will tend to maximize the number of columns judged to be significant, and the pooling down strategy will tend to minimize the number of significant columns. What is the penalty of judging too many columns of being significant or too few?

Alpha and beta mistakes.   In Sec. 3-1 there was reference to improving the quality of the decision of whether to use a new design, process, method, etc., or not to use a new design based on test data. When making this decision, there are four possible outcomes, as shown in Table 3-22. When using the pooling up strategy and judging many columns to be significant, the decision will be to use these factors for further experimentation and perhaps product or process design. The tendency will be to make the alpha mistake more often, thinking that some factor will cause an improvement when in truth that factor will not help. When using the pooling down strategy and judging few columns to be significant, the decision will be to ignore many factors and use only a few for further experimentation and perhaps product or process design. The

---

*G. E. P. Box, W. G. Hunter, J. S. Hunter, *Statistics for Experimenters*. Wiley, New York, 1978, pp. 374–376.

TABLE 3-22 Decision Risks

		The truth about the product	
		There is no improvement	There is some improvement
Decision based on test data	Do *not* use new design	OK	Beta mistake
	Do use new design	Alpha mistake	OK

tendency will be to make the beta mistake more often, thinking that some factor makes no improvement when in truth that factor will help.

From the customer's viewpoint, it is much more serious to make a beta mistake and not take advantage of some amount of improvement offered by a factor. Also, once a factor has been judged to be insignificant, that factor will probably not be included in further rounds of experimentation and the beta mistake will never be exposed. However, if an alpha mistake is made, that factor will be included in further experimentation and the alpha mistake will potentially be exposed. Since it is impossible to make both the alpha and beta mistakes simultaneously, the pooling up strategy should be used which will tend to prevent the beta mistake of ignoring helpful factors. There are books available which explain the alpha-beta risk and the economic consequences of those mistakes.*

Example of pooling error variance estimates. The ANOVA table for the experimental example, Sec. 3-4-6, that has been used predominantly throughout this section would appear as in Table 3-23. In this situation five smaller column effects have been pooled to form an estimate of error variance with 5 degrees of freedom associated with that estimate. As a rule of thumb, pooling up to half of the degrees of freedom in an experiment is advisable. Here that rule was exceeded slightly because two of the column effects were substantially larger than the others. The ANOVA summary table could be rewritten to recognize the pooling as shown in Table 3-24.

3-4-7 Confirmation experiment

The confirmation experiment is the final step in verifying the conclusions from the previous round of experimentation. The optimum condi-

---

*W. J. Diamond, *Practical Experiment Designs for Engineers and Scientists*. Lifetime Learning Publications, Belmont, Calif., 1981, chap. 2.

TABLE 3-23   Pooling of Error Variance

Source	SS	v	V	F
A^*	1.125	1	1.125	
B	21.125	1	21.125	22.83#
$A \times B$	15.125	1	15.125	16.35‡
Col 4*	3.125	1	3.125	
Col 5*	.125	1	.125	
Col 6*	.125	1	.125	
Col 7*	.125	1	.125	
T	40.875	7		
e pooled*	4.625	5	.925	

†At least 90% confidence.
‡At least 95% confidence.
#At least 99% confidence.

tions are set for the significant factors and levels and several tests are made under constant conditions. The average of the confirmation experiment results is compared to the anticipated average (see Sec. 5-4-3), based on the factors and levels tested. The confirmation experiment is a crucial step and should not be omitted.

3-5   Example Experimental Procedure: Popcorn Experiment

This example has been used in seminars to walk people through the process of designing experiments. The scenario is one of developing the process specifications (cooking instructions) to go on a bag of popcorn. The owner of the popcorn company has developed a new hybrid seed which may or may not use the same cooking process recommended for the current seed. One of the processes used by customers to pop corn is the hot oil method, which is addressed in this situation.

TABLE 3-24   Pooling Error Variance ANOVA Summary Table

Source	SS	v	V	F
B	21.125	1	21.125	22.83#
$A \times B$	15.125	1	15.125	16.35‡
e_p	4.625	5	.925	
T	40.875	7		

†At least 90% confidence.
‡At least 95% confidence.
#At least 99% confidence.

3-5-1   Statement of problem and objective of experiment

The problem is to find the process factors which influence popcorn quality characteristics relative to customer requirements. Characteristics such as the number of unpopped kernels in a batch, the fluffiness or volume of popped corn, the color, the taste, and the crispness are typically considered. The objective of the experiment is to find the process conditions which optimize the various quality characteristics to provide improved popping, fluffiness, color, taste, and texture.

3-5-2   Measurement methods

Different measurement methods have to apply for the different performance characteristics that are of interest to customers. The number of unpopped kernels in a batch can be easily measured, but it assumes that in each test there are an equal number of uncooked kernels at the beginning. The fluffiness or volume can be quantified by placing the popped kernels in a measuring cup, which again assumes that an equal number of uncooked kernels were used in each batch. The performance characteristics of color, taste, and texture are somewhat more abstract than the other characteristics. These may be addressed by assigning a numerical color rating, taste rating, and texture rating to each batch.

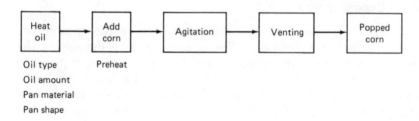

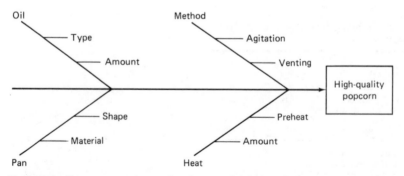

Figure 3-12 Popcorn experiment flowchart and fishbone diagram.

TABLE 3-25 Popcorn Factors and Levels

	Factor	Level 1	Level 2
A	Type of oil	Corn oil	Peanut oil
B	Amount of oil	Low	High
C	Amount of heat	Medium	High
D	Preheat	No	Yes
E	Agitation	No	Yes
F	Venting	No	Yes
G	Pan material	Aluminum	Steel
H	Pan shape	Shallow	Deep

These characteristics can also be treated as attribute data, which is addressed in Chap. 7. Attribute data means that the results fall into classes such as good color, off color, bad color, or good taste and bad taste might be used for taste rating. In any event, a popcorn eating panel will have to be formed to evaluate the test results with respect to color, taste, and texture.

3-5-3 Popcorn process factor and level selection

The factors are selected relative to the popcorn characteristics that customers usually consider in high-quality popcorn. Factors are considered from a process flowchart or fishbone diagram viewpoint to answer the question of what factors may influence the number of unpopped kernels, volume, color, and so on. Factors such as the kind of oil, amount of oil, temperature, preheating, agitation, and venting are typical of suggested factors. The flowchart and fishbone diagram for these factors are shown in Fig. 3-12. For a screening experiment, the factor levels are kept at two levels, a low and a high value. In the case of preheating or venting the level values can be no and yes. The factors and levels are shown in Table 3-25.

3-5-4 Assignment of factors to columns

The factor list is small enough to fit into an L16 OA at a resolution 2 if two levels are used for each factor. The column assignment is easy enough by referring to Table D.2 of the appendix. Factors A to H are assigned to columns 1, 2, 4, 7, 8, 11, 13, and 14. The trial data sheets can then be generated from the factor column assignments. Referring to Table 3-26, the trial data sheet for trial #5 would appear as in Table 3-27.

TABLE 3-26 L16 OA Popcorn Factor Column Assignment

							Popcorn Factors								
	A	B		C			D	E			F		G	H	
							Column no.								
Trial no.	1	2	3	4	5	6	7	8	9	10	11	12	13	14	15
1	1	1	1	1	1	1	1	1	1	1	1	1	1	1	1
2	1	1	1	1	1	1	1	2	2	2	2	2	2	2	2
3	1	1	1	2	2	2	2	1	1	1	1	2	2	2	2
4	1	1	1	2	2	2	2	2	2	2	2	1	1	1	1
5	1	2	2	1	1	2	2	1	1	2	2	1	1	2	2
6	1	2	2	1	1	2	2	2	2	1	1	2	2	1	1
7	1	2	2	2	2	1	1	1	1	2	2	2	2	1	1
8	1	2	2	2	2	1	1	2	2	1	1	1	1	2	2
9	2	1	2	1	2	1	2	1	2	1	2	1	2	1	2
10	2	1	2	1	2	1	2	2	1	2	1	2	1	2	1
11	2	1	2	2	1	2	1	1	2	1	2	2	1	2	1
12	2	1	2	2	1	2	1	2	1	2	1	1	2	1	2
13	2	2	1	1	2	2	1	1	2	2	1	1	2	2	1
14	2	2	1	1	2	2	1	2	1	1	2	2	1	1	2
15	2	2	1	2	1	1	2	1	2	2	1	2	1	1	2
16	2	2	1	2	1	1	2	2	1	1	2	1	2	2	1

3-5-5 Conducting the popcorn experiment

The order of the experiment could be completely randomized with one test per trial. A batch of 200 seeds could be made for made for each trial with the specified trial conditions. For each trial, the number of unpopped kernels, the volume of popcorn, the color rating, the taste rating, and the crispiness rating would be noted.

3-5-6 Popcorn experiment interpretation

Each of the performance characteristics would be analyzed (ANOVA) separately to determine which factors and levels gave the best results for that characteristic. The factors and levels would then be assessed to see if any of the common factors influencing the results were at conflicting levels for the best results. If so, the impact of having the factors at the wrong levels or at some intermediate level as a compromise would have to be assessed.

3-6 Summary

This chapter covered some of the basic weaknesses of typical test strategies and offered some more efficient and powerful alternative strategies in the form of orthogonal arrays. Also covered were the basic steps necessary to designing, conducting, and analyzing OA experiments.

**TABLE 3-27   Trial 5 Data Sheet
(Popcorn Experiment)**

Corn oil
High amount of oil
Medium heat
Preheat oil before adding popcorn
No agitation during popping
Vented during popping
Aluminum pan
Deep pan shape

These fundamentals are necessary for utilizing OAs in some more complex situations which will be covered in the next chapter and some later chapters.

Problems

3-1   If factor A is assigned to column 4 of an L8 OA and factor B assigned to column 6, what column will estimate the $A \times B$ interaction.

3-2   Assign factors A, B, C, D, and E as well as interactions $C \times D$ and $C \times E$ to an OA if all factors are using two levels.

3-3   Assign these factors and interactions to an OA:

A, B, C, D, E, and F (two levels)
$A \times B$ $B \times C$ $C \times E$
$A \times C$ $B \times D$ $D \times E$
$A \times D$ $B \times E$
     $B \times F$

Multiple-Level Experiments

4-1 Necessity for Multiple-Level Experiments

This chapter deals with modification of standard two-level arrays to handle factors of three or four levels. Three-level arrays will handle three-level factors without modification, but a mixture of two-, three-, and four-level factors requires special treatment of standard two-level arrays.

As was mentioned in the previous chapter, there are occasions when discrete factors will be assigned to an experiment and more than two levels will be required. There are situations involving continuous variables that may make three levels very useful; for example, the steering wheel position in an automobile may be set in the left, center, or right positions. However, it is still recommended for an initial experimental investigation to begin with two levels whenever possible (see Sec. 3-4-2). Once significant factors are identified, multiple levels can be used for estimating nonlinear responses.

4-2 Conversion from Two to Four Levels

A two-level array can be converted to contain some four-level columns very simply. The concept for conversion depends upon the requirement

TABLE 4-1   L8 Columns 1, 2, and 3

Trial no.	Column no.		
	1	2	3
1	1	1	1
2	1	1	1
3	1	2	2
4	1	2	2
5	2	1	2
6	2	1	2
7	2	2	1
8	2	2	1

for a four-level factor to have 3 degrees of freedom associated with it. Recalling that a two-level OA has 1 degree of freedom allocated to each column, then three columns must be merged to provide sufficient degrees of freedom for a four-level factor. It is recommended to merge three columns which are mutually interactive, such as columns 1, 2, and 3. The merging of mutually interactive columns minimizes confounding of interactions as much as possible.

If columns 1, 2, and 3 are merged in an L8 to form a single four-level column, the eight trials need to have a particular level assigned to maintain the orthogonality of the array. A recommended technique for accomplishing this is as follows. The first three columns of an L8 array appear in Table 4-1. If any two columns are studied, one will notice there are four combinations possible of levels 1 and 2.

$$1\ 1 \qquad 1\ 2 \qquad 2\ 1 \qquad 2\ 2$$

Level 1 for the four-level factor can correspond to the 1 1 condition, level 2 to the 1 2 condition, level 3 to the 2 1 condition, and level 4 to the 2 2 condition. Using columns 1 and 2 in Table 4-1, the corresponding levels would appear as in Table 4-2. The three merged columns provide the necessary degrees of freedom, and any two of those columns provide an orthogonal pattern for the four levels. The four levels need not be in consecutive order with respect to the trial numbers; in this case the foremost requirement is the balanced number of two trials at each level.

The resultant L8 array modified to include 1 four-level factor and up to 4 two-level factors is shown in Table 4-3. The columns have been renumbered in this array. Many other combinations of the four-level factor and the remaining two-level columns are possible. Note that it is not possible to include 2 four-level factors in an L8 array without causing some confounding of portions of the factor effects. Also, recall that an L8 linear graph has interaction lines that join at a common point. If three

TABLE 4-2 Four-Level Factor Arrangement

Trial no.	Column no. 1	2	3	Four-level factor
1	1	1	1	1
2	1	1	1	1
3	1	2	2	2
4	1	2	2	2
5	2	1	2	3
6	2	1	2	3
7	2	2	1	4
8	2	2	1	4

mutually interactive columns are selected, another set cannot be found that does not have some column in common.

4-3 ANOVA for Four-Level Factors

The ANOVA formulas used in Chap. 2 apply in this instance. The four-level factor requires the general formula. Following is a demonstration of the method. The property of orthogonality is undisturbed. If factors A (four level), B, C, D, and E are assigned to an L8 OA and responses are as shown in Table 4-4, then the analysis for the factors (columns) and total variation is

$$SS_A = \frac{A_1^2}{n_{A_1}} + \frac{A_2^2}{n_{A_2}} + \frac{A_3^2}{n_{A_3}} + \frac{A_4^2}{n_{A_4}} - \frac{T^2}{N} \quad \text{(see Sec. 2-3-3)}$$

$$= \frac{8^2}{2} + \frac{11^2}{2} + \frac{17^2}{2} + \frac{20^2}{2} - \frac{56^2}{8}$$

$$= 45.0$$

TABLE 4-3 L8 OA Modified for
a Four-Level Factor

Trial no.	Column no. 1	2	3	4	5
1	1	1	1	1	1
2	1	2	2	2	2
3	2	1	1	2	2
4	2	2	2	1	1
5	3	1	2	1	2
6	3	2	1	2	1
7	4	1	2	2	1
8	4	2	1	1	2

TABLE 4-4 Four-Level Experiment Analysis

Trial no.	Factors					y data
	A	B	C	D	E	
	Column no.					
	1	2	3	4	5	
1	1	1	1	1	1	2
2	1	2	2	2	2	6
3	2	1	1	2	2	4
4	2	2	2	1	1	7
5	3	1	2	1	2	7
6	3	2	1	2	1	10
7	4	1	2	2	1	8
8	4	2	1	1	2	12

$$SS_B = \frac{(B_1 - B_2)^2}{N} = \frac{(35 - 21)^2}{8} = 24.5$$

$$SS_C = \frac{(28 - 28)^2}{8} = 0.0$$

$$SS_D = \frac{(28 - 28)^2}{8} = 0.0$$

$$SS_E = \frac{(29 - 27)^2}{8} = 0.5$$

$$SS_T = 2^2 + 6^2 + 4^2 + 7^2 + 7^2 + 10^2 + 8^2 + 12^2 - \frac{56^2}{8}$$

$$= 70.0$$

Note that the sums of squares for all the columns add up to the total sum of squares, so the orthogonality of the experiment is still present.

As another demonstration of the preservation of the orthogonality, but not used in a typical experimental analysis, is the fact that the sums of squares for the original three columns that were merged, when added together, equal the sum of squares for factor A.

$$SS_A = SS_{col\,1} + SS_{col\,2} + SS_{col\,3}$$

$$SS_{col\,1} = \frac{(1_1 - 1_2)^2}{N} = \frac{(19 - 37)^2}{8} = 40.50$$

$$SS_{col\,2} = \frac{(2_1 - 2_2)^2}{N} = \frac{(25 - 31)^2}{8} = 4.50$$

$$SS_{col\,3} = \frac{(3_1 - 3_2)^2}{N} = \frac{(28 - 28)^2}{8} = 0.00$$

$$45.00 = 40.50 + 4.50 + 0.00$$

The ANOVA table for the example appears in Table 4-5. A problem is created when interpreting the F ratios relative to each other when two factors have an unequal number of levels. The F ratios seem to indicate that the factor B effect is much larger than the factor A effect, but what does a plot of the data indicate? Figure 4-1 shows the four-level factor data plotted. The F ratios reflect the relative magnitudes of the incremental effects of both factor A and B. Factor A has three increments adding up to the total factor A effect, and factor B has one increment. However, since the total span of level A_1 to A_4 is available from the product or process viewpoint, the total effect of factor A is available also. Care must be exercised when interpreting F values when multiple-level factors and two-level factors are mixed in an experiment. Don't mistake the magnitude of the F ratios to indicate the magnitude of the overall effect. Statistically, the F ratio indicates the presence of a factorial effect, not necessarily the size of the effect. Plots of statistically significant factors will indicate the magnitudes of the effects by the slopes; the higher the slope, the greater the factor effect. The F ratio basically indicates which factors and interactions should be plotted.

Other interpretations that may be made of the plot depend upon the type of response; lower is better (LB), nominal is best (NB), or higher is better (HB). If an HB characteristic, the treatment condition of A_4B_2 is the best within the investigated experimental space. If an LB characteristic, the treatment condition of A_1B_1 is the best within the experimental space. Either one of these characteristics would suggest further experimentation to increase or decrease the average result for an HB or LB characteristic, respectively. An NB characteristic could be interpo-

TABLE 4-5 ANOVA for OA Modified to Four Levels

Source	SS	ν	V	F
A	45.0	3	15.0	90#
B	24.5	1	24.5	147#
$C^\dagger$	0.0	1	0.0	
$D^\dagger$	0.0	1	0.0	
$E^\dagger$	0.5	1	0.5	
T	70.0	7		
e pooled†	0.5	3	0.167	

†At least 90% confidence.
‡At least 95% confidence.
$^\#$At least 99% confidence.

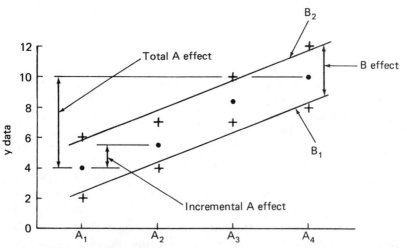

Figure 4-1 Four-level factor plot.

lated fairly accurately to estimate a particular treatment condition that would provide the proper average.

4-4 Polynomial Decomposition

When specific conditions are met in an experiment, the factor sum of squares may be decomposed in even smaller sources. If the factor evaluated is of a continuous nature having equal increments between levels and an equal sample size at all levels, then polynomial decomposition is possible using a simple method. Table 4-6 shows the various polynomial effects that may be estimated. The linear and quadratic effect are frequently found, although a factor with a pure cubic effect is rarely seen.

Another example will be used to demonstrate the polynomial decomposition method. Table 4-7 shows some possible results for an L8 OA modified to a four-level first column arrangement (remaining columns are omitted for clarity). The sum of squares for factor A is

$$\text{SS}_A = \frac{A_1^2}{n_{A_1}} + \frac{A_2^2}{n_{A_2}} + \frac{A_3^2}{n_{A_3}} + \frac{A_4^2}{n_{A_4}} - \frac{T^2}{N}$$

TABLE 4-6 Polynomial Effects for Increasing Levels

No. of levels	Polynomial effect
2	Linear
3	Quadratic
4	Cubic

TABLE 4-7 Four-Level Experimental Data

Trial no.	A	y data
1	1	4
2	1	5
3	2	8
4	2	9
5	3	10
6	3	9
7	4	5
8	4	4

$$SS_A = \frac{9^2}{2} + \frac{17^2}{2} + \frac{19^2}{2} + \frac{9^2}{2} - \frac{54^2}{8}$$

$$= 41.5$$

The SS_A value has 3 degrees of freedom associated with it. Those 3 degrees of freedom can be broken down into the components shown in Table 4-8. Each of the degrees of freedom may be used to estimate a polynomial effect. The number of polynomial effects that may be estimated is equal to $k - 1$, where k equals the number of levels, or to the degrees of freedom for that factor. A two-level factor may only have the linear effect estimated, a three-level factor may have the linear and quadratic effects estimated, etc.

The method for calculating the sum of squares for the various polynomial effects is based on the same equation.

$$SS_{A \text{ polynomial effect}} = \frac{(W_1 A_1 + W_2 A_2 + \cdots + W_k A_k)^2}{W_T R}$$

Table D.3 in Appendix D contains the W coefficients that are appropriate for a particular polynomial effect. The table is set up for from two-level to five-level factors. The R value is equal to the number of samples under a level.

TABLE 4-8 Components of Four-Level Factor Variation

Source	ν
$SS_{A \text{ linear}}$	1
$SS_{A \text{ quadratic}}$	1
$SS_{A \text{ cubic}}$	1
SS_A	3

In this example the W coefficients are taken from the four-level column. The various polynomial effects are then calculated

$$SS_{A \text{ linear}} = \frac{[-3(9) - 1(17) + 1(19) + 3(9)]^2}{20(2)} = 0.1$$

$$SS_{A \text{ quadratic}} = \frac{[1(9) - 1(17) - 1(19) + 1(9)]^2}{4(2)} = 40.5$$

$$SS_{A \text{ cubic}} = \frac{[-1(9) + 3(17) - 3(19) + 1(9)]^2}{20(2)} = 0.9$$

$$SS_A = SS_{A \text{ linear}} + SS_{A \text{ quadratic}} + SS_{A \text{ cubic}} = 41.5$$

This example demonstrates how the sums of squares can be decomposed into polynomial effects. Also, the different polynomial effects can be F-tested to determine if any are statistically significant. In this example, the quadratic effect is much larger than any other. Figure 4-2 shows the relative magnitudes. The data is sloped slightly up to the right, which is proportional to the linear effect of factor A. The graph has one maximum and is greatly curved, which is proportional to the quadratic effect of factor A. And the slight rate of change of the curvature is proportional to the cubic effect of factor A. If the A_3 data points were values of 8 and 9, there would be no linear or cubic effect but purely a quadratic effect present. A parabola (quadratic equation) would fit the averages of the data at the four levels exactly. The two data points at each level of factor A imply the error variation that is present around each of the level averages. This is one method to determine if the data such as in Sec. 2-3-13 follows a curved or straight line. The smaller insignificant

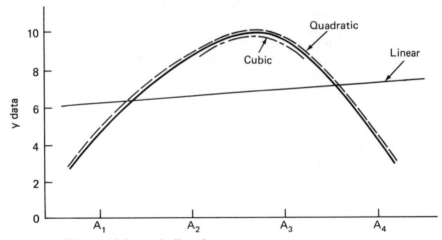

Figure 4-2 Polynomial factorial effect plot.

polynomial effects can be pooled for error estimation with other factors if necessary.

Remember, the polynomial decomposition depends on

1. Continuous variables
2. Equal increments between levels
3. Equal sample sizes at all levels

4-5 Multiple-Level Factor Interactions

Occasionally, it may be necessary to include an interaction of a two-level factor and a four-level factor. The interaction can obviously be evaluated in an OA, but some columns must be allocated for that interaction. The four-level factor was created by merging three mutually interactive columns. A separate two-level factor could be assigned to a column, say column 4, in the two-level array. Then the interaction would be contained in the columns indicated in Fig. 4-3. Recall that the interaction is going to require the product of the degrees of freedom of the interacting factors. In this case the interaction has 3 degrees of freedom associated with it; adding the sums of squares for columns 5, 6, and 7 would estimate the interaction effect. The portion represented in any column is unpredictable; only the total can be used for the sum of squares due to the interaction.

4-6 Dummy Treatment for Three-Level Factors

The dummy treatment method is based on the four-level method when modifying a basic two-level OA. Three mutually interacting columns are merged and the pattern for the four levels determined. The dummy

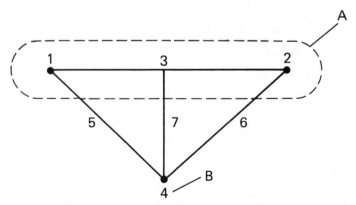

Figure 4-3 Four-level factor interaction linear graph.

treatment accommodates three-level factors by using only three of the four possible levels for the factor and the indicated fourth level simply repeating one of the previous three levels. Any one of the three levels for the factor can be repeated as shown in Table 4-9. Any one of the three levels can be repeated and maintain orthogonality, so whichever is easiest, cheapest, or makes more sense should be repeated.

4-7   ANOVA for a Dummy Treatment

The column to which factor A is assigned has 3 degrees of freedom associated with it. Factor A, however, has only 2 degrees of freedom, so the remaining unassigned degree of freedom must be associated with error. Because of this dual source of variation, the ANOVA must be performed in a slightly different manner. An example will facilitate understanding of the ANOVA method. Table 4-10 shows possible experimental results. The formula for sum of squares due to factor A must use the general form

$$SS_A = \frac{A_1^2}{n_{A_1}} + \frac{A_2^2}{n_{A_2}} + \frac{A_3^2}{n_{A_3}} - \frac{T^2}{N}$$

The level symbols for factor A_1 and $A_{1'}$, both indicate the same test condition with respect to factor A. Therefore,

$$SS_A = \frac{(A_1 + A_{1'})^2}{n_{A_1} + n_{A_{1'}}} + \frac{A_2^2}{n_{A_2}} + \frac{A_3^2}{n_{A_3}} - \frac{T^2}{N}$$

$$= \frac{(15 + 17)^2}{2 + 2} + \frac{12^2}{2} + \frac{8^2}{2} - \frac{52^2}{8} = 22.00$$

TABLE 4-9   L8 Dummy Treatment for Factor A

	Four levels	Three-level options		
	Merged columns			
Trial no.	1 2 3		1 2 3	
1	1	1	1	1
2	1	1	1	1
3	2	2	2	2
4	2	2	2	2
5	3	3	3	3
6	3	3	3	3
7	4	1'	2'	3'
8	4	1'	2'	3'

NOTE: ' indicates the repeated level.

TABLE 4-10 Experimental Results Using Dummy Treatment

	A	B	C	D	E	
			Column no.			
Trial no.	1	2	3	4	5	y (data)
1	1	1	1	1	1	10
2	1	2	2	2	2	5
3	2	1	1	2	2	8
4	2	2	2	1	1	4
5	3	1	2	1	2	1
6	3	2	1	2	1	7
7	1'	1	2	2	1	6
8	1'	2	1	1	2	11

The variation due to error can be treated in the same manner as a two-level factor, that is, a comparison of the averages under the 1 and 1' conditions. The denominator of the fraction is the total number of tests involved in that comparison.

$$SS_e = \frac{(A_1 - A_{1'})^2}{n_{A_1} + n_{A_{1'}}}$$

$$= \frac{(15 - 17)^2}{4} = 1.00$$

As an exercise, the remaining sources of variation and total variation should be calculated for the example. The total variation can still be accounted for in the columns. Table 4-11 summarizes the ANOVA calculations for this example.

A plot of the significant factors in Fig. 4-4 shows how the dummy treatment is used and causes an unequal number of samples in the

TABLE 4-11 ANOVA Summary for Dummy Treatment

Source	SS	ν	V	F
A	22.00	2	11.00	22.0#
$B^\dagger$	0.50	1	0.50	
C	50.00	1	50.00	100.0#
$D^\dagger$	0.00	1	0.00	
$E^\dagger$	0.50	1	0.50	
$e^\dagger$	1.00	1	1.00	
T	74.00	7		
$e^\dagger$ pooled	2.00	4	0.50	

†At least 90% confidence.
‡At least 95% confidence.
#At least 99% confidence.

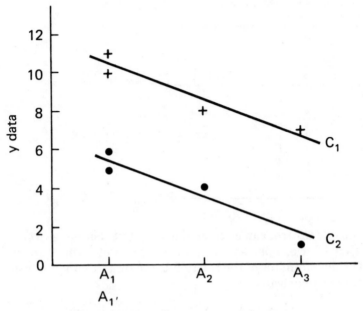

Figure 4-4 Dummy treatment plot.

different test conditions. The orthogonality of the experiment is still intact, however, because the total of the sums of squares for the merged columns now equals the sum of squares for factor A plus the error sum of squares.

$$SS_A + SS_e = SS_{col\,1} + SS_{col\,2} + SS_{col\,3}$$

4-8 Summary

This chapter discussed the methods for modifying a standard two-level OA to handle a three- or four-level factor, and the subsequent accommodations required to complete the ANOVA. The next chapter covers some additional interpretation techniques to supplement the typical ANOVA table and plotting methods discussed up to this point.

Problems

4-1   If columns 2, 4, and 6 are merged to form a four-level column, what will the level designation be for trials 1 to 8 for an L8 OA?

4-2   Columns 2, 4, and 6 are merged to provide a four-level arrangement for factor A. Calculate SS_A for the data indicated.

Trial no.	1	2	3	4	5	6	7	8
Factor A level	1	2	4	3	1	2	4	3
y data	3	6	10	8	4	5	9	8

4-3 Calculate SS_B and SS_e for the dummy treatment data.

Trial no.	1	2	3	4	5	6	7	8
Factor B level	1	2	2′	3	1	2	2′	3
y data	3	0	1	6	4	1	1	7

4-4 Decompose polynomially this experimental data.

Trial no.	1	2	3	4	5	6	7	8
Factor C level	1	1	2	2	3	3	4	4
y data	4	5	2	1	6	6	10	9

Interpretation of Experimental Results

5-1 Interpretation Methods

Once an experiment has been conducted, an ANOVA completed, the significant factors and/or interactions identified, and plots made of the various treatment conditions, some amount of interpretation has already been done. However, other pieces of information are useful for the experimenter and since analysis of data is inexpensive compared to the cost of conducting the test, it is good judgment to completely analyze the data.

Several items will be discussed in this chapter:

1. Percent contribution
2. Estimating the mean
3. Confidence interval around the estimated mean
4. Omega transformation of data
5. Other interpretation methods

5-2   Percent Contribution

The portion of the total variation observed in an experiment attributed to each significant factor and/or interaction is reflected in the percent contribution. The percent contribution is a function of the sums of squares for each significant item. The percent contribution indicates the relative power of a factor and/or interaction to reduce variation. If the factor and/or interaction levels were controlled precisely, then the total variation could be reduced by the amount indicated by the percent contribution.

When using an experiment to resolve a production problem, the total variation observed in an experiment should represent more than 75% of the variation observed in production. This is an indicator, but not a certainty, that the appropriate factors have been included in the screening experiment. If the range of data in the experiment is very small relative to the range of data in production, then, even if factors are significant, they may not be powerful enough to control the problem and other factors will need to be investigated.

Although it has not been mentioned in the ANOVA sections, the variance due to a factor or interaction contains some amount due to error.[*][†] An equation that states this for factor A, for example, is

$$V_A = V'_A + V_e$$

V'_A is the expected amount of variance due solely to factor A. Solving for V'_A,

$$V'_A = V_A - V_e \tag{5-1}$$

Recall that the definition of variance for factor A is

$$V_A = \frac{SS_A}{\nu_A}$$

Then

$$V'_A = \frac{SS'_A}{\nu_A}$$

Substituting into Eq. (5-1),

$$\frac{SS'_A}{\nu_A} = \frac{SS_A}{\nu_A} - V_e$$

---

*C. R. Hicks, *Fundamental Concepts in the Design of Experiments*. New York, Holt, Rinehart, & Winston, 1982.

†P. W. M. John, *Statistical Design and Analysis of Experiments*. New York, Macmillan, 1971.

Solving for SS$'_A$,

$$SS'_A = SS_A - (V_e) (v_A)$$

SS$'_A$ is the expected sum of squares due to factor A, and the percent P of the contribution to the total variation can now be calculated.

$$P = \frac{SS'_A}{SS_T} \times 100$$

This example used factor A, but any factor or interaction can be substituted as long as the appropriate sums of squares and associated degrees of freedom are used.

Two more columns, SS$'$ and P, can be added to complete an ANOVA table. Since some portion of the sums of squares for a factor and/or interaction was subtracted out because of error, this amount must be added to the error sum of squares in order that the total sum of squares is unchanged. The total percent contribution must add to 100 percent. The percent contribution due to error provides an estimate of the adequacy of the experiment. If the percent contribution due to error (unknown and uncontrolled factors) is low, 15% or less, then it is assumed that no important factors were omitted from the experiment. If it is a high value, 50% or more, then some important factors were definitely omitted, conditions were not precisely controlled, or measurement error was excessive.

The ANOVA table for the example in Sec. 4-3 would now appear as in Table 5-1. Note that in this case the interpretation problem of the ranking of the F ratios is rectified by observing the percent contribution. Here, the percent contribution indicates that factor A contributes the most toward the variation observed in the experiment. This agrees with the plot in Fig. 4-1, even though the F ratio indicates that factor B is more significant. See the discussion in Sec. 4-3 concerning the relative incremental effects of factors A and B.

TABLE 5-1   ANOVA including Percent Contribution

Source	SS	v	V	F	SS$'$	P
A	45.00	3	15.00	90.0*	44.50	63.60
B	24.50	1	24.50	147.0*	24.33	34.80
$C^‡$	0.0	1	0.00			
$D^‡$	0.0	1	0.00			
$E^‡$	0.5	1	0.50			
T	70.00	7			70.00	100.00
e^* pooled	0.50	3	0.167		1.17	1.60

†At least 90% confidence.
‡At least 95% confidence.
*At least 99% confidence.

5-3   Estimating the Mean

Typically, the experimenter would like to obtain some particular response from a product or process; a higher average response is better (HB), a nominal value is best (NB), or a lower average response is better (LB). Depending on the characteristic, different treatment combinations will be chosen to obtain satisfactory results.

When an experiment has been conducted and the optimum treatment condition within the experiment determined, one of two possibilities exists:

1. The prescribed combination of factor levels is identical to one of those in the experiment
2. The prescribed combination of factor levels was not included in the experiment (the lower the resolution, smaller fraction of a full factorial experiment, the more likely this is to occur)

If situation 1 exists, then the most direct way to estimate the mean for that treatment condition is to average all the results for the trials which are set at those particular levels. If situation 2 exists, then a more indirect route will have to be taken to predict the average for that treatment condition. This method may also be used in situation 1, which will utilize more of the data to estimate the average.

The procedure depends upon the additivity of the factorial effects. If one factorial effect can be added to another to accurately predict the result, then good additivity exists. If an interaction exists, then the additivity between those factors is poor. The additivity of the interaction and other noninteracting factors, however, may be good.

A plot of two noninteracting factors is shown in Fig. 5-1. Geometrically, the midpoint on the B_1 line represents $\overline{B}_1$, the average of all the data under the B_1 condition. The same applies for the B_2 condition. $\overline{A}_1$ may be found midway between the B_1 and B_2 condition when factor A is at the first level. The same applies for $\overline{A}_2$. If these four points are connected by two line segments as shown, $\overline{A}_1$–$\overline{A}_2$ and $\overline{B}_1$–$\overline{B}_2$, then the intersection represents $\overline{T}$, the average of the entire experimental results.

Assume the A_2B_2 treatment condition is to be estimated. Then $(\overline{A}_2 - \overline{T})$ represents the A_2 effect to change the average from $\overline{T}$ to $\overline{A}_2$ and $(\overline{B}_2 - \overline{T})$ represents the B_2 effect to change the average from $\overline{T}$ to $\overline{B}_2$. Since there is no interaction the additivity is good.

$$\hat{\mu}_{A_2B_2} = \overline{T} + (\overline{A}_2 - \overline{T}) + (\overline{B}_2 - \overline{T}) = \overline{A}_2 + \overline{B}_2 - \overline{T}$$

This is a very simple situation, but any number of factors can be combined if their additivity is good. The coefficient on the $\overline{T}$ term is one less

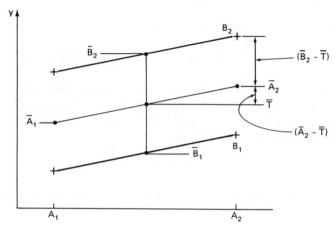

Figure 5-1 Noninteracting factors plot.

than the number of items added to estimate the mean. In this case two items were added, so the coefficient is equal to one. Another example provides an estimate of the $B_2C_1F_2G_2$ condition:

$$\hat{\mu}_{B_2C_1F_2G_2} = \overline{B}_2 + \overline{C}_1 + \overline{F}_2 + \overline{G}_2 - 3\overline{T}$$

If the characteristic is an HB characteristic and:

$$\begin{aligned}
\overline{B}_1 &= 1.30 & \overline{B}_2 &= 1.50 \\
\overline{C}_1 &= 2.30 & \overline{C}_2 &= 0.50 \\
\overline{F}_1 &= 0.90 & \overline{F}_2 &= 1.90 \\
\overline{G}_1 &= 1.00 & \overline{G}_2 &= 1.80 \\
\overline{T} &= 1.40 & &
\end{aligned}$$

$$\begin{aligned}
\hat{\mu}_{B_2C_1F_2G_2} &= 1.50 + 2.30 + 1.90 + 1.80 - 3(1.40) \\
&= 3.30
\end{aligned}$$

Note that the average of the $B_2C_1F_2G_2$ condition is higher than any of the individual factor averages because of the additivity of the effects.

When an interaction exists, the additivity is poor between the factors, especially when the interaction effect is larger than any of the main effects. To obtain the best estimate of a mean when an interaction is present, the trials that include that treatment condition should be averaged. Referring to Fig. 2-13, the hardness level is an HB characteristic, so the condition A_2B_2 is the best in the experiment. The effect of A_2B_2 condition to change the average from $\overline{T}$ is represented by

$$\overline{A_2B_2} - \overline{T}_2$$

When two factors are interacting then

$$\overline{A}_2 + \overline{B}_2 - \overline{T} \ne \overline{A_2B_2} \tag{5-2}$$

The interaction effect is the difference between the left and right sides of Eq. (5-2).

If two other factor levels, C_1 and D_1, happen to also improve the hardness, how can the total effect be estimated? By considering the interaction as one item which has good additivity to other noninteracting items will allow an estimate to be made

$$\hat{\mu}_{A_2B_2C_1D_1} = \overline{A_2B_2} + \overline{C}_1 + \overline{D}_1 - 2\overline{T}$$

Here the poor additivity of the interacting factors is avoided when other noninteracting items are added. If

$$\overline{A_2B_2} = 9.5 \qquad \overline{C}_1 = 7.2 \qquad \overline{D}_1 = 7.5 \qquad \overline{T} = 6.875$$

Then

$$\hat{\mu}_{A_2B_2C_1D_1} = 9.5 + 7.2 + 7.5 - 2(6.875) = 10.45$$

Again, any number of items can be combined when the additivity is good and the coefficient on $\overline{T}$ is one less than the number of items combined.

5-4   Confidence Interval around the Estimated Mean

The estimate of the mean $\hat{\mu}$ is only a point estimate based on the averages of results obtained from the experiment. Statistically this provides a 50% chance of the true average being greater than $\hat{\mu}$ and a 50% chance of the true average being less than $\hat{\mu}$. The experimenter would prefer to have a range of values within which the true average would be expected to fall with some confidence. The confidence interval is a maximum and minimum value between which the true average should fall at some stated percentage of confidence.

Confidence, in the statistical sense, means there is some chance of a mistake. For instance, the confidence that a number of 1, 2, 3, 4, or 5 could be rolled on a standard die is 5/6 or 83%. It is possible to roll a 6 so there is some risk of being wrong. When stating a confidence value for a confidence interval, experimenters are simply stacking the odds in their favor that the true average will fall between the stated limits. A high confidence may be chosen to reduce risk, but a wider confidence interval will result, lowering the chance of the true average being outside the stated limits.

There are three different types of confidence intervals (CIs) that Taguchi uses depending on the purpose of the estimate.

1. Around the average for a particular treatment condition in the existing experiment

2. Around the estimated average of a treatment condition predicted from the experiment

3. Around the estimated average of a treatment condition used in a confirmation experiment to verify predictions

5-4-1 CI_1 for existing experimental treatment condition

This method of calculating a CI is the traditional statistical approach.

$$CI_1 = \sqrt{\frac{F_{\alpha;1;\nu_2}V_e}{n}}$$

where $F_{\alpha;1;\nu_2} = F$ ratio required for:

   α = risk   Confidence = $1 -$ risk
   $\nu_1 = 1$
   ν_2 = degrees of freedom for error = ν_e
   V_e = error variance
   n = number of tests under that condition

The F ratio is determined from the same F tables used in ANOVA. The 1 degree of freedom for the numerator associated with the mean that is being estimated will always be a value of 1 for a confidence interval. The degrees of freedom for the denominator are the degrees of freedom ν_e associated with the pooled error variance V_e of the experiment.

The CI is used in this manner:

$$\hat{\mu}_{A_1} = \overline{A}_1 \pm CI_1$$

or

$$\overline{A}_1 - CI_1 \leqslant \hat{\mu}_{A_1} \leqslant \overline{A}_1 + CI_1$$

Both of these statements are made at some chosen confidence level.

As an example of the use of a CI, the data for Fig. 2-12 has a 90% confidence interval of

$$CI_1 = \sqrt{\frac{F_{\alpha;1;\nu_e}V_e}{n}}$$

where $F_{.10;1;9} = F$ ratio for 90% confidence = 3.36
   $V_e = .052$
   $n = 4$

$$CI_1 = \sqrt{\frac{3.36\,(.052)}{4}} = .21$$

$$\overline{A}_1 = 1.050 \qquad \overline{A}_2 = 1.475 \qquad \overline{A}_3 = 2.050$$

$$1.050 - .21 < \hat{\mu}_{A_1} < 1.050 + .21$$

$$0.840 < \hat{\mu}_{A_1} < 1.260$$

$$1.265 < \hat{\mu}_{A_2} < 1.685$$

$$1.840 < \hat{\mu}_{A_3} < 2.260$$

As shown in Fig. 5-2, the CIs do not overlap each other at the 90% confidence level, indicating at least 90% confidence that the averages are different from each other. Also, a straight line passes through all of the CIs, indicating that the performance does not have a nonlinear component of any significance statistically. This method is complementary to polynomial decomposition for interpretation of product or process performance.

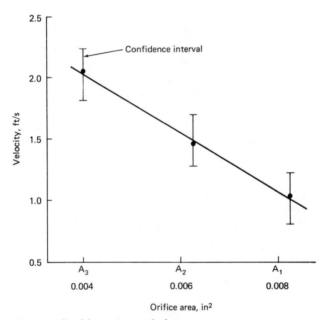

Figure 5-2 Confidence interval plot.

5-4-2 CI_2 for predicted treatment condition

The CI calculation is modified:

$$CI_1 = \sqrt{\frac{F_{\alpha;1;v_e}V_e}{n_{\text{eff}}}}$$

$$n_{\text{eff}} = \frac{N}{1 + \left[\begin{array}{c}\text{total degrees of freedom associated} \\ \text{with items used in } \hat{\mu} \text{ estimate}\end{array}\right]}$$

For example, if an L16 had been used for the experiment with one test (repetition) per trial and

$$\hat{\mu}_{A_2B_2C_1D_2} = \overline{A_2B_2} + \overline{C}_1 + \overline{D}_2 - 2\overline{T}$$

Then $N = 16$ and

$$n_{\text{eff}} = \frac{16}{1 + 1 + 1 + 1} = 4$$

One will note, a four-factor assignment to an L16 does not have the $A_2B_2C_1D_2$ combination. The Taguchi method of using n_{eff} gives credit for the data points used in calculating $\overline{A_2B_2}$, $\overline{C}_1$, and $\overline{D}_2$. Again,

$$\hat{\mu}_{A_2B_2C_1D_2} = (\overline{A_2B_2} + \overline{C}_1 + \overline{D}_2 - 2\overline{T}) \pm CI_2$$

Another example of the n_{eff} calculation is

$$\hat{\mu}_{A_3B_3C_2} = \overline{A}_3 + \overline{B}_3 + \overline{C}_2 - 2\overline{T}$$

$$v_A = 2 \qquad v_B = 2 \qquad v_C = 1$$

$$n_{\text{eff}} = \frac{N}{1 + 2 + 2 + 1} = \frac{N}{6}$$

5-4-3 CI_3 for predicting a confirmation experiment

A confirmation experiment is used to verify that the factors and levels chosen from an experiment cause a product or process to behave in a certain fashion. A selected number of tests are run under constant, specified conditions to observe results that, the experimenter hopes, are close to the predicted value.

The difference between CI_2 and CI_3 is that CI_2 is for the entire population, i.e., all parts ever made under the specified conditions, and CI_3 is for only a sample group made under the specified conditions.

Because of the smaller sample size relative to the entire population, CI_3 must be slightly wider and is modified.

$$CI_3 = \sqrt{F_{\alpha;1;\nu_e} V_e[(1/n_{\text{eff}}) + (1/r)]}$$

r = sample size for confirmation experiment, $r \neq 0$

As r approaches infinity, i.e., the entire population, the value $1/r$ approaches zero and $CI_2 = CI_3$. As r approaches 1, the CI becomes wider.

The confirmation experiment is highly recommended to verify the experimental conclusions and is interpreted in this manner. If the average of the results of the confirmation experiment is within the limits of the confidence interval, then the experimenter believes the significant factors as well as the appropriate levels for obtaining the desired result were properly chosen. If the average of the results of the confirmation experiment is outside the limits of the CI, then the experimenter has selected the wrong factors and/or levels to control the results at a desired value or has excessive measurement error, necessitating further experimentation. Other factors and other levels will have to be included in further rounds of experimentation. The confirmation experiment is the last step in an investigation to verify the understanding of what makes the product or process function properly.

5-5   Omega Transformation

The omega transformation of data is used in another situation where poor additivity can happen. That is when the data is in percentage values such as percent yield, percent loss, or percent defective. As these values approach 100% or 0%, the additivity may be poor; when the $\hat{\mu}$ value is calculated, a value greater than 100% or less than 0% can be obtained.

To use the omega conversion method:

1. Convert data percent values to db values using the omega tables or formula
2. Use $\hat{\mu}$ equation to estimate the mean with substituted omega values
3. Convert the obtained db value back to the percent value using the omega tables or formula

The omega tables are in Table D.5 of the appendix.

In a fully saturated resolution 1 experiment several factors may contribute something to reducing the percent defectives in a process. One or two of the factors probably contribute the most, but any help in reducing loss is useful. For example (percent loss is an LB characteristic),

$$\overline{A}_1 = 5\% \qquad \overline{C}_1 = 10\% \qquad \overline{D}_1 = 3\% \qquad \overline{F}_1 = 8\%$$

$$\overline{A}_2 = 9\% \qquad \overline{C}_2 = 4\% \qquad \overline{D}_2 = 11\% \qquad \overline{F}_2 = 6\%$$

$$\overline{T} = 7\%$$

Then

$$\hat{\mu}_{A_1C_2D_1F_2} = \overline{A}_1 + \overline{C}_2 + \overline{D}_1 + \overline{F}_2 - 3\overline{T}$$

$$= 5 + 4 + 3 + 6 - 3(7) = -3\%$$

Apply the omega transformation from Appendix D.

%	db
5	-12.787
4	-13.801
3	-15.096
6	-11.949
7	-11.233

$$\hat{\mu}_{A_1C_2D_1F_2} = -12.787 - 13.801 - 15.096 - 11.949 + 3(11.233)$$
$$= -19.934 \text{ db}$$

Converting the db value back to percentage makes the estimate equal to approximately 1 percent. The omega transformation formula is

$$\Omega \text{ (db)} = 10 \log [p/(1 - p)]$$

where p = percent values $(0 < p < 1)$.

The omega transformation converts fractions between 0 and 1 to values between minus infinity and plus infinity. This transformation is most useful because percentage values are very small or very large. When percentages are between 20 and 80%, the additivity is generally good.

5-6   Other Interpretation Methods

There are several alternative interpretation methods that are available because of the structure of the orthogonal arrays. These should be used as complementary to ANOVA and not as a total replacement for more sophisticated statistical methods, especially when computer software is available to do analysis of Taguchi OAs.

5-6-1 Observation method

This method is suggested as a preliminary interpretation method, and it is possible to use it when the response is an LB or HB characteristic. The experiment in Table 3-6 to reduce leaks in engine water pumps is a very good example to use for the observation method.

The seven factors mentioned in the table were assigned to an L8 OA and eight different assemblies were rated for leakage. The leak performance was monitored on a semi-variable scale with 0 representing a no-leak condition and a 5 representing the worst leak condition which had ever been observed. The experimental results are shown in Table 5-2. Trials 4 and 7 have the most desirable results (leak rating is an LB characteristic) from a customer viewpoint, but what have those trials in common that might provide better pump performance? When looking at column 1, factor A, one can see that trial 4 has level A_1 and trial 7 has level A_2. Since good results were obtained in either case, it is unlikely that factor A contributes anything toward a successful result. Columns 2, 5, and 7 have common levels for trials 4 and 7 (B_2, E_2, and G_1). Presumably, one or all of these factors at these levels contribute toward successful results. At this point any interaction that may exist is irrelevant; this particular combination seems to provide good results. So instead of having to control one factor, three will have to be controlled. This would be contingent, of course, on further interpretation of this data, which may help isolate the significant factors. A confirmation test should be run to verify the appropriateness of the factors and levels. Follow-up experimentation should be used to isolate the significant factors and any interactions that might exist.

TABLE 5-2 Engine Water Pump Leak Test

			Factors				
	A	B	C	D	E	F	G
				Column no.				
Trial no.	1	2	3	4	5	6	7	Leak rating data
1	1	1	1	1	1	1	1	4
2	1	1	1	2	2	2	2	3
3	1	2	2	1	1	2	2	1
4	1	②	2	2	②	1	①	⓪
5	2	1	2	1	2	1	2	2
6	2	1	2	2	1	2	1	4
7	2	②	1	1	②	2	①	⓪
8	2	2	1	2	1	1	2	1

TABLE 5-3   Engine Water Pump Leak Test (Ranked Results)

			Factors				
	A	B	C	D	E	F	G
				Column no.			
Trial no.	1	2	3	4	5	6	7	Leak rating data
4	1	2	2	2	2	1	1	0
7	2	2	1	1	2	2	1	0
3	1	2	2	1	1	2	2	1
8	2	2	1	2	1	1	2	1
5	2	1	2	1	2	1	2	2
2	1	1	1	2	2	2	2	3
1	1	1	1	1	1	1	1	4
6	2	1	2	2	1	2	1	4

5-6-2   Ranking method

An extension of the observation method is to rank all of the results, from the best to the worst, and the corresponding trial conditions as shown in Table 5-3. With this method consistency of the levels at the ends of the goodness to badness scale are of interest. Here there is a very strong relationship with all of the B_2 levels falling toward the goodness end and all of the B_1 levels on the badness end. This is the only factor with this strong a relationship. Factor E has moderate strength since two 2s are at the good end and two 1s at the bad end. Note, also, the relationship for factor E remains intact, within each of the factor B levels indicating secondary strength to factor B. A plot of the results would show very little, if any, interactive effect between factors B and E, with the B effect being much stronger than the E effect. The new gasket design in conjunction with a smooth pump housing finish should provide a leakproof unit.

Both the observation and ranking method have an interesting property when using two-level OAs as the experimental basis; that is, to divide the results into groups which are consistent within the group. If one factor is really the important factor, then there will tend to be two groups of data, one associated with the low level and one associated with the high level. If two factors are really important, then there will tend to be four groups; two trials in an L8 may be good results or four trials in an L16 may be good results. If some odd number of trials appears in a group, some note should be made of this. Because of this two-level design, the groups should be made up of an even number of tests.

TABLE 5-4   Engine Water Pump Leak Test (C-E Method)

	Factors							
	A	B	C	D	E	F	G	
	Column no.							
Trial no.	1	2	3	4	5	6	7	Leak rating data
1	1	1	1	1	1	1	1	4
2	1	1	1	2	2	2	2	3
3	1	2	2	1	1	2	2	1
4	1	2	2	2	2	1	1	0
5	2	1	2	1	2	1	2	2
6	2	1	2	2	1	2	1	4
7	2	2	1	1	2	2	1	0
8	2	2	1	2	1	1	2	1
Sum_1	8	13	8	7	10	7	8	
Sum_2	7	2	7	8	5	8	7	
Difference	−1	−11	−1	1	−5	1	−1	

5-6-3   Column effects method

This approach is used by Taguchi as a simplified ANOVA to point out columns which have large influences on the response. Using the same leak experiment and analyzing by the column effects (C-E) method is shown in Table 5-4.

The sum of the data associated with the first level is subtracted from the sum of the data associated with the second level of each column. The magnitudes of the differences are compared to each other to find the relatively larger effects. One will recognize the level difference for each column as the fundamental value in the sum of squares formula for ANOVA; however, no statistical criterion is determined to allow us to make a decision as to which columns are significant. The relative magnitudes (the plus or minus sign shows positive or negative correlation, respectively) indicate the relative power of the factors in effecting the results. Again, factor B has the largest effect, factor E the second largest effect, and all other factors have a weak effect.

The observation, ranking, and column effects methods all indicate the same factors and relationship for improving the sealability of the engine water pump. No statistical decision criteria were used for any of these cursory analysis techniques. They should be used in conjunction with ANOVA for comprehensive evaluation.

5-6-4   Plotting analysis

The plotting of experimental results should be done in as many ways as are meaningful

1. By levels of significant factors

2. By levels of combinations of significant factors to assess interaction possibilities

3. By order of actual test

Plotting in order of actual test will reveal any strong trends, up or down, that may have existed during the execution of the experiment. The trends would have to be caused by some unknown, uncontrolled factor(s) that varied during the experiment. Hopefully, the data will show no correlation (no trend up or down) with the test order, which would disturb the interpretation of results. If a trend up or down is detected, the data should be corrected to account for the trend and analyzed after correction.

To correct the data for a trend so that it appears as a horizontal pattern is very simple. A trend line is drawn through the data with respect to actual test order. The data values should have the difference between the trend line and $\overline{T}$ subtracted from the data in the area where the trend line is above $\overline{T}$. The difference between the trend line and $\overline{T}$ should be added to the data in the area where the trend line is below $\overline{T}$. This removes the trend effect from the data which reduces error variation and makes ANOVA more sensitive to differences in averages that may exist.

If a trend should exist, it is recommended that the experimenter pursue the factor(s) that may have produced the trend.

5-6-5 Secondary rounds of experimentation

The initial screening experiment is intended to detect the factors which produce the desired effect. However, within the levels that were chosen for the experiment, the optimum results may not have been achieved and further experimentation may be necessary to move toward the optimum. Further experimentation should be done using significant factors from the screening experiment but with new levels that are anticipated to improve performance.

Depending on the type of characteristic (LB, NB, or HB), different strategies have to be used. Figure 5-3 shows the possibilities that might exist for a single-factor situation or a two-factor situation.

If one factor possesses the ability to influence the product or process response, then achieving a higher, nominal, or lower value is relatively easy. Appropriately extrapolating or interpolating along the line of best fit to the results will allow the experimenter to use a certain level of the factor of concern to obtain a certain result (refer to graph A of Fig. 5-3).

Interactions between factors are not as easily addressed and depend on the type of characteristic as indicated in graphs B, C, and D in Fig.

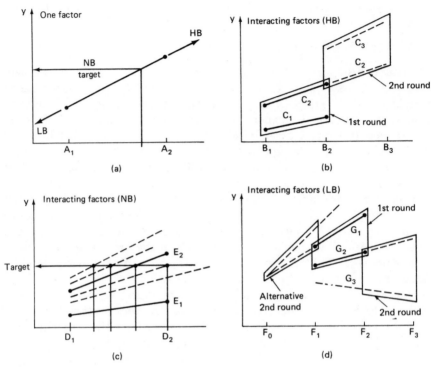

Figure 5-3 Secondary rounds of experimentation. (*a*) One factor; (*b*) interacting factors (HB); (*c*) interacting factors (NB); (*d*) interacting factors (LB).

5-3. The HB and LB characteristics are very similar in approach. New levels for the factors of concern are chosen as a way to extrapolate into yet uninvestigated territory. The extrapolated lines need the verification of the secondary round of experimentation. This round of experimentation should include one of the levels of the primary round as a way to verify the repeatability of the experiment. If the average of the duplicated condition of the second round is within the confidence interval of the first round, then this is an indicator, but not a guarantee, that the controlling factors are in the experiment.

If some of the experimental results are higher than the target value and some are lower for a NB characteristic, then several combinations of the two factors will provide the desired nominal value. Economics relevant to these factors and their levels should be considered in choosing the factor levels. If the target value is greater than all of the interaction treatment condition averages, then the same approach as shown for an HB characteristic should be used. If the target value is less than all of the interaction treatment condition averages, then the same approach as shown for an LB characteristic should be used.

5-7 Summary

This chapter covered some additional interpretational methods to ANOVA to aid the experimenter in fully understanding the product or process. Some very simple methods were introduced which can help the speedy intepretation of experimental results. All analytical tools at the disposal of an experimenter should be considered when the experiment has been conducted. The next chapter provides a few special designs for some common problems encountered in experimentation.

Special Designs

Some situations occur in typical product or process development projects which are not accommodated by standard two- or three-level OAs. The following sections describe three special OA arrangements and one adaptation of an OA.

6-1 Nested Experiments

This situation arises when experimenting with discrete (noncontinuous) factors. The use of a discrete factor may cause each half of a two-level OA to have entirely different kinds of test conditions. One example proposed two kinds of retention methods, factor A, for a plug inserted into the end of a valve. One retention method was to simply press-fit the plug into place and the other was to heat-stake the plug after press-fit assembly. The factors that apply to the heat-staking process are not at all relevant when only press fitting the plug into position. If the retention method is assigned to the first column of an OA, then the upper half of the experiment could be related to heat staking and the lower half to not heat staking. If a heat-staking factor, such as the amount of heat-staking time, factor B, is assigned to the second column, levels 1 and 2 in the lower half of the experiment are meaningless with respect to not heat staking (see Table 6-1). This condition requires the use of a nested

TABLE 6-1 Nested Factor and Error Column Assignment

	Factors		
	A	B	
	Column no.		
Trial no.	1	2	3
1	1	1	$\not 1$
2	1	1	$\not 1$
3	1	2	$\not 2$
4	1	2	$\not 2$
5	2	1	$\not 2$
6	2	1	$\not 2$
7	2	2	$\not 1$
8	2	2	$\not 1$

experiment to accommodate the discrete factor(s). The time factor is nested within the retention method factor.

The nested experiment requires a special modification to accommodate the discrete factor(s). The first step is to identify the primary factors in the experiment. These are the factors which are varied over the entire experiment. The nesting factor is considered a primary factor. The retention method is a nesting factor. Second, the nested factors (secondary factors) are assigned to columns which interact with the nesting factor column assignment. The degrees of freedom are allocated in the manner shown in Fig. 6-1a. An error estimate will come from the half of the experiment when the parts are press fit and time of heat staking is meaningless. Table 6-1 shows this assignment.

Trials 1 to 4 in column 1 relate to heat staking and trials 5 through 8 relate to press fitting. Trials 1 to 4 in column 2 relate to low and high time of heat staking and trials 5 to 8 relate to low and high values for error estimation. Since there are 2 degrees of freedom effectively present in column 2, the third column is omitted to donate the necessary degree of freedom to column 2.

Other nested factors can be joined to nesting factor A through interacting columns, and a degree of freedom used for error once again. Factor C, such as temperature of heat staking, could be nested in factor A. Factor C would have to be assigned to a separate column; consequently, the interaction column for $A \times C$ would be omitted to contribute its degree of freedom for error in the factor C column. Table 6-2 contains a potential list of factors for this experiment. Figure 6-1b shows the column assignment for three nested factors and a factor which is not nested in factor A. Factor E applies only when the parts are not heat-staked, so a more compact OA arrangement may be used. Since the lower half of column 2 was unassigned (error estimate), two factors can be

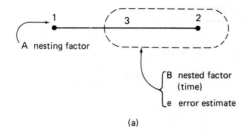

(a)

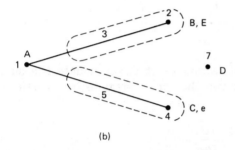

(b)

Figure 6-1 Nested factors linear graph.

assigned to this column. The upper half of column 2 applies to factor B and the lower half applies to factor E. Each factor will be evaluated in the half of the experiment to which it pertains. Table 6-3 shows the complete factor assignment. Primary factors A and D are varied over the entire experiment, whereas secondary factors B, C, and E are effective in only one-half of the experiment.

If the press-out force were measured for each of the assemblies and had values as in Table 6-3, the analysis for the primary factors would be

$$SS_A = \frac{(A_1 - A_2)^2}{N} = \frac{(36 - 20)^2}{8} = 32.00$$

$$SS_D = \frac{(D_1 - D_2)^2}{N} = \frac{(17 - 39)^2}{8} = 60.50$$

TABLE 6-2 Heat Staking Factors

Factors		Level 1	Level 2
A	(retention method)	Heat-staked	Not heat-staked
B	(stake time)	Low	High
C	(stake temperature)	Low	High
D	(press fit amount)	Low	High
E	(press fit depth)	Flush	Below

TABLE 6-3 Nested Factor Column Assignment

		Factors				
	A	B,E	C,e		D	e
			Column no.			Force,
Trial no.	1	2	3		4	5	lb
1	1	1	} B	1	} C	1	1	4
2	1	1		2		2	2	10
3	1	2		1		2	2	14
4	1	2		2		1	1	8
5	2	1	} E	1	} e	1	2	3
6	2	1		2		2	1	8
7	2	2		1		2	1	7
8	2	2		2		1	2	2

NOTE: 1 lb = .454 kg.

The analysis for the nested factors must accommodate the smaller sample size within the nest and the error estimate in a portion of a column:

$$SS_B = \frac{(B_1 - B_2)^2}{(n_{B_1} + n_{B_2})}$$

$$= \frac{(14 - 22)^2}{2 + 2}$$

$$= 16.00$$

$$SS_C = \frac{(C_1 - C_2)^2}{(n_{C_1} + n_{C_2})}$$

$$= \frac{(18 - 18)^2}{2 + 2}$$

$$= 0.00$$

$$SS_E = \frac{(E_1 - E_2)^2}{(n_{E_1} + n_{E_2})}$$

$$= \frac{(11 - 9)^2}{2 + 2}$$

$$= 1.00$$

$$SS_{e\ col\ 3} = \frac{(3_1 - 3_2)^2}{(n_{31} + n_{32})}$$

$$SS_{e \text{ col } 3} = \frac{(10 - 10)^2}{(2 + 2)} = 0.00$$

$$SS_{e \text{ col } 5} = \frac{(5_1 - 5_2)^2}{N}$$

$$= \frac{(27 - 29)^2}{8} = 0.50$$

Table 6-4 shows the ANOVA summary for this particular set of experimental conditions and data.

Figure 6-2 shows individual plots for the three significant factors. Since press-out force is an HB characteristic, the treatment condition of $A_1B_2D_2$ is thought to provide the maximum value of the force.

Taguchi uses the pooling up strategy for F-testing the remaining factors in the experiment. Traditionally, however, statisticians recommend using the error variance within a nest (group) to F-test the factors within the nested factor. It is very possible the two retention methods do not have equal variance, which violates one of the assumptions of ANOVA. Taguchi is trying to identify from a practical viewpoint which factors affect performance and uses the F test as a reference value. The confirmation experiment is used to validate the combined effects of the factors selected.

6-2 Combined Factors

6-2-1 Background

Sometimes the factors selected for evaluation include a mixture of two-level and three-level factors. A two-level array can be used in conjunc-

TABLE 6-4 ANOVA Summary for a Nested Experiment

Source	SS	ν	V	F
A	32.0	1	32.0	84.21*
B	16.0	1	16.0	42.11*
$C^\dagger$	0.0	1	0.0	
D	60.5	1	60.5	159.21*
$E^\dagger$	1.0	1	1.0	
$e^\dagger$	0.5	2	0.25	
T	110.0	7		
$e_p^\dagger$	1.5	4	0.38	

†At least 90% confidence.
‡At least 95% confidence.
*At least 99% confidence.

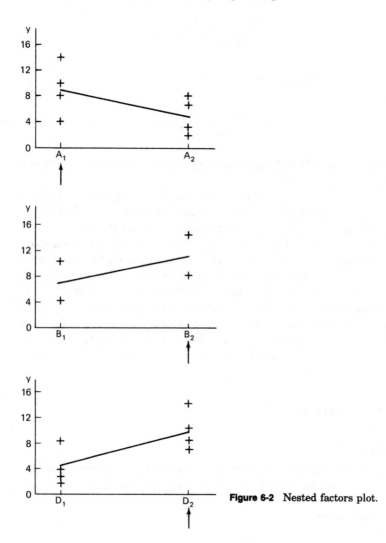

Figure 6-2 Nested factors plot.

tion with the dummy treatment method for each three-level factor treated in this fashion. A three-level array may also be used and dummy treat the two-level factors, which requires a degree of freedom for error for every two-level factor. If the factor list is at the limit of degrees of freedom for the OA, then all of the factors may not be allocated. Rather than eliminate factors, it is better to include them in an experiment of some form.

6-2-2 Combined factors method

The combined factors approach treats a pair of two-level factors as a three-level factor. However, the estimate of any interaction between those factors must be sacrificed. The four possible combinations of two factors at two levels are, for example,

$$A_1B_1 \qquad A_2B_1 \qquad A_1B_2 \qquad A_2B_2$$

If the A_1B_1 condition, A_2B_2 condition, and one of the remaining two conditions is selected, then two comparisons can be made to estimate the factor A and B effects. Assuming these test conditions:

Combined factor level		
Level 1	Level 2	Level 3
A_1B_1	A_2B_1	A_2B_2

Comparing level 1 to level 2 provides an estimate of factor A's effect (factor B is held constant at B_1) and level 2 compared to level 3 provides an estimate of factor B's effect (factor A is held constant at A_2). Figure 6-3 shows how this test arrangement can estimate the factor effects when these particular treatment conditions are selected. If there happens to be an interaction between factors A and B, then one of the factor effects will be overestimated and one will be underestimated. However, no knowledge of which factor is overestimated and which is underestimated

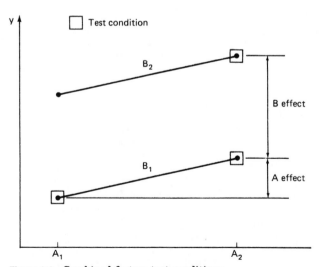

Figure 6-3 Combined factors test conditions.

TABLE 6-5   Combined Factors Column Assignment

	Factors				
	AB	C	D	E	
	Column no.				
Trial no.	1	2	3	4	Data
1	1	1	1	1	2
2	1	2	2	2	6
3	1	3	3	3	11
4	2	1	2	3	6
5	2	2	3	1	10
6	2	3	1	2	1
7	3	1	3	2	14
8	3	2	1	3	6
9	3	3	2	1	10

will be available; it will not be known whether the nonparallel lines for an interaction would be converging on the left or right side of the graph.

These three test conditions can be assigned to one column in a three-level OA as a combined factor AB. Because the A_1B_2 condition is not tested, the orthogonality of factors A and B is lost. An example L9 OA and data demonstrate the impact of the lost orthogonality. Table 6-5 shows an L9 combined factor column assignment and test results. The sum of squares calculations for the total variation and variation due to factors are

$$SS_T = 2^2 + 6^2 + \cdots + 10^2 - \frac{66^2}{9} = 146.00$$

$$SS_{AB} = \frac{(AB)_1^2}{n_{(AB)_1}} + \frac{(AB)_2^2}{n_{(AB)_2}} + \frac{(AB)_3^2}{n_{(AB)_3}} - \frac{T^2}{N}$$

$$SS_{AB} = \frac{19^2}{3} + \frac{17^2}{3} + \frac{30^2}{3} - \frac{66^2}{9} = 32.67$$

$$SS_C = \frac{22^2}{3} + \frac{22^2}{3} + \frac{22^2}{3} - \frac{66^2}{9} = 0.0$$

$$SS_D = \frac{9^2}{3} + \frac{22^2}{3} + \frac{35^2}{3} - \frac{66^2}{9} = 112.67$$

$$SS_E = \frac{22^2}{3} + \frac{21^2}{3} + \frac{23^2}{3} - \frac{66^2}{9} = 0.66$$

The total orthogonality has not been destroyed:

$$SS_T = SS_{AB} + SS_C + SS_D + SS_E$$

$$146.00 = 32.67 + 0.0 + 112.67 + 0.67$$

The sum of squares can be calculated for the comparison of $(AB)_1$ to $(AB)_2$ by applying the specific two-level formula

$$SS_A = \frac{[(AB)_1 - (AB)_2]^2}{n_{(AB)_1} + n_{(AB)_2}}$$

$$= \frac{[19 - 17]^2}{3 + 3} = 0.67$$

Similarly, for factor B,

$$SS_B = \frac{[(AB)_2 - (AB)_3]^2}{n_{(AB)_2} + n_{(AB)_3}}$$

$$= \frac{[17 - 30]^2}{3 + 3} = 28.17$$

As a demonstration of the lost orthogonality between factors A and B, the following inequality can be written.

$$SS_{AB} \neq SS_A + SS_B$$

$$32.67 \neq 0.67 + 28.17$$

Therefore, the orthogonality between factors A and B is lost.

In this situation, the factor A effect is very small and the factor B effect is very large; hence, factor B must be the significant factor of the two. If this is true, then the combined factors treatment is equivalent to a dummy treatment of factor B. With a dummy treatment the orthogonality of the entire experiment is preserved. Level $(AB)_2$ can be thought of as the dummy level for factor B which would make that level the $B_{1'}$. Calculating the sum of squares for factor B as if it were dummy-treated, which assumes the factor A effect is zero, would give these results:

$$SS_B = \frac{(B_1 + B_{1'})^2}{n_{B_1} + n_{B_{1'}}} + \frac{B_2^2}{n_{B_2}} - \frac{T^2}{N}$$

$$= \frac{(19 + 17)^2}{3 + 3} + \frac{30^2}{3} - \frac{66^2}{9} = 32.00$$

$$SS_e = \frac{(B_1 - B_{1'})^2}{n_{B_1} + n_{B_{1'}}}$$

$$SS_e = \frac{(19 - 17)^2}{3 + 3} = 0.67$$

So, therefore,

$$SS_{AB} = SS_B + SS_e$$

$$32.67 = 32.00 + 0.67$$

Since the entire experiment is orthogonal for a dummy treatment,

$$SS_T = SS_A + SS_B + SS_C + SS_D + SS_E + SS_e$$

$$SS_A = 0.0$$

$$146.00 = 0.0 + 32.00 + 0.0 + 112.67 + 0.66 + 0.67$$

This can only be done if the factor A or B effect is very small relative to the other effects.

6-3 Idle Column Method

The idle column method can be used to fit several three-level factors into a two-level OA with many two-level factors. This method also partially destroys orthogonality but it eliminates the need for dummy treatment and the subsequent provision for an error degree of freedom for each three-level factor. For two or more three-level factors, the idle column method saves experimental trials.

The idle column linear graph assignment appears in Fig. 6-4, which provides the 2 degrees of freedom for each three-level factor A and B. Mutually interactive columns are chosen with the idle column serving as a common column.

The levels for the three-level factors are dictated by the levels indicated in the idle column. When the idle column indicates level 1, then levels 1 and 2 are assigned for the three-level factor. When the idle column indicates level 2, then levels 2 and 3 are assigned for the three-level factor. The OA is set up for one factor as shown in Table 6-6 for an L8

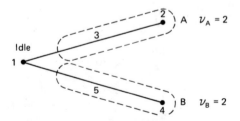

Figure 6-4 Idle column linear graph.

TABLE 6-6   Idle Column Assignment for L8 OA

	Factors		
	Idle	A	
	Column no.		
Trial no.	1	2	3
1	1	1	~~1~~
2	1	1	~~1~~
3	1	2	~~2~~
4	1	2	~~2~~
5	2	~~1~~ 2	~~2~~
6	2	~~1~~ 2	~~2~~
7	2	~~2~~ 3	~~1~~
8	2	~~2~~ 3	~~1~~

OA. Levels 2 and 3 are substituted for 1 and 2, respectively, when the idle column indicates level 2. One of the two columns which were merged for the three-level factor must be eliminated from the OA; in this case column 3 was removed.

The analysis of a factor using an idle column assignment is done in two parts. $SS_{A_{(1-2)}}$ and $SS_{A_{(2-3)}}$ are added together to obtain the total SS_A effect. However, because of using the idle column method, part of factor A's effect is in the idle column and is not able to be estimated. When several factors are associated with the idle column, a mixture of parts of each of the factorial effects plus error effects are in the idle column. Because of this confounding of factorial and error effects, a factor should not be assigned to the idle column unless it is a nuisance variable that is to be blocked in the experiment.

If 2 three-level factors (A and B) and 2 two-level factors (C and D) were assigned to an L8 OA using column 1 as the idle column, the assignment would appear as in Table 6-7. Original columns 3 and 5 were removed from the OA. The sum of squares calculations are

$$SS_T = 82.0$$

$$SS_{A_{(1-2)}} = \frac{(9 - 19)^2}{4} = 25.0$$

$$SS_{A_{(2-2)}} = \frac{(19 - 21)^2}{4} = 1.0$$

$$SS_{B_{(1-2)}} = \frac{(16 - 12)^2}{4} = 4.0$$

$$SS_{B_{(2-3)}} = \frac{(20 - 20)^2}{4} = 0.0$$

$$SS_C = \frac{(26 - 42)^2}{8} = 32.0$$

$$SS_D = \frac{(32 - 36)^2}{8} = 2.0$$

$$SS_{idle} = \frac{(28 - 40)^2}{8} = 18.0$$

The ANOVA summary is shown in Table 6-8.

Here factor $A_{(1-2)}$ is significant but $A_{(2-3)}$ is not, which indicates some difference in averages from condition A_1 to A_2, but no detectable difference in averages from A_2 to A_3.

$$I_1 \begin{cases} \overline{A}_1 = \dfrac{9}{2} = 4.5 \\[2ex] \overline{A}_2 = \dfrac{19}{2} = 9.5 \end{cases}$$

$$I_2 \begin{cases} \overline{A}_2 = \dfrac{19}{2} = 9.5 \\[2ex] \overline{A}_3 = \dfrac{21}{2} = 10.5 \end{cases}$$

$$\frac{38}{4} = 9.5 = \overline{A}_2$$

The comparison of levels A_1 to A_2 involved a total of four samples, as did the comparison of A_2 to A_3 when the idle column indicated levels 1 and 2, respectively. Therefore, the confidence interval calculation would be

$$CI_1 = \sqrt{\frac{F_{.10;1;4} \, V_e}{n_{eff}}} = \sqrt{\frac{2.84(1.75)}{2}} = 1.58$$

TABLE 6-7 Idle Column Experiment Example

	Idle	A	B	C	D	
			Factors			
			Column no.			
Trial no.	1	2	3	4	5	Data
1	1	1	1	1	1	3
2	1	1	2	2	2	6
3	1	2	1	2	2	13
4	1	2	2	1	1	6
5	2	2	2	1	2	8
6	2	2	3	2	1	11
7	2	3	2	2	1	12
8	2	3	3	1	2	9

TABLE 6-8   ANOVA Summary for Idle Column Experiment

Source	SS	v	V	F
Idle	18.0	1	18.0	10.3[‡]
$A_{(1-2)}$	25.0	1	25.0	14.3[‡]
$A^{\dagger}_{(2-3)}$	1.0	1	1.0	
$B^{\dagger}_{(1-2)}$	4.0	1	4.0	
$B^{\dagger}_{(2-3)}$	0.0	1	0.0	
C	32.0	1	18.3	
$D^{\dagger}$	2.0	1	2.0	
T	82.0	7		
$e^{\dagger}_p$	7.0	4	1.75	

[†]At least 90% confidence.
[‡]At least 95% confidence.
[#]At least 99% confidence.

A plot of the comparisons is shown in Fig. 6-5. If this were an HB characteristic, then level A_3 would have the highest average and could be used, depending on economics, even though statistically no different than A_2. If the A_3 condition is a cost penalty, then perhaps A_2 should be used to obtain nearly the same performance at a lower cost. The estimate of the mean would be

$$\hat{\mu}_{A_3 C_2} = \overline{A}_3 + \overline{C}_2 - \overline{T}$$

The confidence interval for the population would be CI_2 which depends upon n_{eff}.

$$n_{\text{eff}} = \frac{N}{1 + \left[\begin{array}{c} \text{total degrees of freedom associated} \\ \text{with items used in } \hat{\mu} \text{ estimate} \end{array} \right]}$$

$$n_{\text{eff}} = \frac{8}{1 + 2 + 1} = 2$$

6-4   Summary of Multiple Levels

Four different methods of accommodating multiple levels have been reviewed in the text:

1. *Merging columns.*   A four-level factor can be fit into a two-level OA (see Sec. 4-2).
2. *Dummy treatment.*   A three-level factor can be fit into a two-level OA (see Sec. 4-6).
3. *Combination method.*   Two-level factors can be fit into a three-level OA (see Sec. 6-2).
4. *Idle column method.*   Many three-level factors can be fit into a two-level OA (see Sec. 6-3).

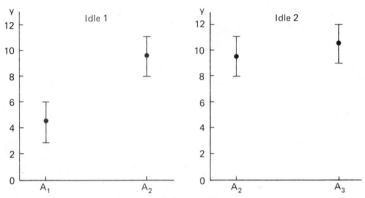

Figure 6-5 Idle column comparison of averages.

The first two methods do not disturb the orthogonality of the entire experiment. The last two methods cause a loss of orthogonality among the factors treated by those methods and subsequently the loss of an accurate estimate of the independent factorial effects.

6-5 Component Search

This method is an adaptation of the use of two-level OAs when the reassembly and retest of a product are possible. A test that is destructive in nature is very difficult if not impossible to do in this manner. A component (or subassembly) search is possible if a product is tested and some units are found to be relatively good performers and others poor performers. The experiment can be completed with as little as one good unit and one bad unit but with some limitations which will be discussed later.

The component search method begins by identifying the number of major components or subassemblies within a total assembly that might affect the performance characteristic of interest. The number of identified components dictates the size of the OA; again, one degree of freedom will have to be allocated to each component to be evaluated.

Let's assume that a certain percentage of automatic transmission hydraulic pump assemblies are failing the pressure and flow test at the final test for pumps. If seven major components are identified, then an L8 OA can be the basis for the experiment. The next step would be to collect eight pump assemblies that are poor pressure and/or flow performers and eight pumps that are good performers. It may take some time to collect the eight poor assemblies from a production line unless the problem is severe. The hypothesis behind this experiment is that there are one or more reasons which consistently cause the differences between the good and poor performers.

The pumps are disassembled into groups of common components with all the good components segregated from the poor components. For instance, two groups of eight pump drive gears each should exist when the disassembly is complete; eight gears from the good units and eight gears from the bad units. The pump assemblies are rebuilt into sixteen complete assemblies by using the L8 OA as shown in Table 6-9. In this OA level 1 can symbolize a component from a poor performing unit and level 2 can symbolize a component from a good performing unit (or vice versa). Two pump assemblies can be built in the combination indicated in each trial. The components should be randomly selected from the poor group of components when a level 1 is shown in the OA and randomly selected from the good group when a level 2 is shown.

The ANOVA of the data should indicate a propensity for the data to fall into a pattern that will match the column pattern for the component(s) contributing to the problem. Hopefully, error variance will be small, which indicates that disassembly and reassembly do not affect the pump performance. If the error variance is large, then excess measurement error and/or assembly sensitivity is indicated.

If only four good units and four poor units are used in this experiment, then the effect of disassembly and reassembly can be evaluated by completely disassembling, reassembling, and retesting the pumps. The two tests for each trial should be very consistent unless reassembly affects performance or measurement error exists since the identical components are retested.

This method can be used with as little as one good unit and one poor unit by using the parts to build the particular combination specified for each trial. If this assembly were torn down, completely reassembled, and retested, then an estimate of error variance would be obtained.

An example of using the latter technique involved a problem with manual transmission production. Squealing noise in third range was

TABLE 6-9 Component Search Experiment

		Components							
	A	B	C	D	E	F	G		
			Column no.					Pump flow	
Trial no.	1	2	3	4	5	6	7	y_1	y_2
1	1	1	1	1	1	1	1	*	*
2	1	1	1	2	2	2	2	*	*
3	1	2	2	1	1	2	2	*	*
4	1	2	2	2	2	1	1	*	*
5	2	1	2	1	2	1	2	*	*
6	2	1	2	2	1	2	1	*	*
7	2	2	1	1	2	2	1	*	*
8	2	2	1	2	1	1	2	*	*

*Flow data.

noted in some gearboxes on the final test stands. One noisy and one quiet gearbox were selected from production. Several components related to third range were chosen as items of interest in the investigation. Six items were selected which allowed the use of an L8 OA with one column available for the rest of the transmission assembly from each noisy and quiet gearbox. The factors for the experiment are listed in Table 6-10. The assignment of factors to an L8 OA and the subsequent test results are shown in Table 6-11. The entire experiment was completed in six hours, since the response was a performance characteristic rather than a durability characteristic, which may have taken a long time.

In this experiment, no method was used to quantify the noise level; only a notation was made as to whether the noise was present or not. The pattern of the noise response matched exactly the pattern of levels for column 7, the third range gear. Some test engineers were skeptical of the conclusions that the gear was the only problem and ran several other combinations of parts in an attempt to support their theory. However, the noise remained associated with the third range gear only.

The noisy and quiet gears were then compared dimensionally to isolate any physical differences that existed between the parts. The component search method identifies components but not the root cause of the problem. At this writing, those differences are yet to be determined. Once a list of differences is established, an experiment should be designed to investigate those factors at the levels that were found in the noisy and quiet gears.

In this case study the cause of noise was associated with one particular component. Had the noise been associated with more than one component or an interaction between components, the noise level would have more variation than just the two values of noisy and quiet. If this were the situation, then a more sophisticated measurement system for noise would probably have to be used for meaningful results.

A test engineer in the noise lab commented when the test was complete, "We've learned more in six hours using this method than we have in months on previous problems using other methods." A sparse amount

TABLE 6-10 Noisy Transmission Component Search

Factors		Level 1	Level 2
A	Sleeve	Noisy	Quiet
B	Fork	Noisy	Quiet
C	Transmission assembly	Noisy	Quiet
D	Input shaft	Noisy	Quiet
E	Bearing	Noisy	Quiet
F	Blocker	Noisy	Quiet
G	Gear	Noisy	Quiet

TABLE 6-11 Manual Transmission Third Range Noise Experiment

Trial no.	A	B	C	D	E	F	G	Noise
	1	2	3	4	5	6	7	
1	1	1	1	1	1	1	1	S
2	1	1	1	2	2	2	2	Q
3	1	2	2	1	1	2	2	Q
4	1	2	2	2	2	1	1	S
5	2	1	2	1	2	1	2	Q
6	2	1	2	2	1	2	1	S
7	2	2	1	1	2	2	1	S
8	2	2	1	2	1	1	2	Q

Factors *A B C D E F G* (Column no.)

NOTE: S = squeal; Q = quiet.

of data that is collected strategically is worth more than a large amount of data with a poor collection strategy.

6-6 Summary

This chapter addressed some special experimental design situations typically faced by the engineer or scientist. The nested factors section illustrated the method of handling a situation of a factor which causes two completely different treatments in halves of an experiment. Methods of accommodating two-level factors in three-level OAs and three-level factors in two-level OAs were discussed. A summary of the possible OA modifications is covered in Sec. 6-4. An interesting application of OAs is the component search method when reassembly and retest of a device is possible. All of the data analysis up to this point in the text has addressed the use of variable data. The next chapter covers the use and analysis of attribute date when applied to the OA experimental design.

Chapter

7

Attribute Data

7-1 Introduction to Attribute Data

All of the discussions and analyses of experimental situations so far have dealt with variable data as the performance characteristic (observation or response) of interest. Variable data is a continuous form, which means that an infinite number of values can occur anywhere between very low values and very high values. The discrimination between any two experimental results depends only on the precision of the measurement system. Examples of variable data are temperature, pressure, flow rate, efficiency, power, weight, length, velocity, volume, and time. The temperature produced by a chemical reaction might be a performance characteristic of interest (instant cold packs) which could technically assume any value between −460°F and an extremely large positive value of temperature.

Attribute data, on the other hand, is a discontinuous form, which means the experimental results can only be discrete values such as good and bad or off and on or 0 and 1. A single-pole, single-throw switch can assume either the off position or the on position; there is no such thing as half off, quarter off, or quarter on position. In a production process a defect of a product is either present or not present. Numerically, this situation can be represented by assigning a value of 1 to a part with a defect and a 0 to a part without a defect (or vice versa for analytical purposes).

Two-class attribute data provides much less discrimination than variable data. When a part is classified as bad, no measure of how bad is provided. When a part is classified as good, no measure of how good is provided. Because of this reduced discrimination, many pieces of attribute data are required to provide the equivalent information as one piece of variable data. Variable data should be used whenever possible due to the greatly reduced sample size.

Two-class attribute data can be of two types. One kind is a situation where the total number of tests, the number of good results, and the number of bad results are known. For example, a certain number of engine cyclinders are ignited during a test run and the number of occurrences of normal combustion and the number of occurrences of detonation recorded. Another kind of attribute data is where the total number of tests is known and only the number of one class of occurrence may be known. For example, an experiment to grow apples would recognize the number of apples on a tree. The number of nonapples would not or could not be known.

To improve the discrimination power of attribute data more classes may be used. For instance, if parts were rated on the severity of a defect present, then several classes may be used:

Class no.	Description
1	No defect
2	Mild defect
3	Moderate defect
4	Severe defect

As the number of classes increases, the data becomes semivariable in nature, which adds power of discrimination. In this instance the class number has some engineering or scientific meaning; the higher the class number, the more severe the defect. In some classified data the frequency of occurrence in a particular class may not have such meaning; the number of deaths due to heart attack by city block, for instance. Here the block number does not necessarily indicate a more or less severe condition.

Multiple-class, more than two classes, data has other advantages over two-class data which concerns the ability to detect a shift in average or to detect an increase or decrease in variability. The greater the number of classes, the more a frequency distribution by class will resemble a histogram and provide a clue to the shape of the distribution. With only two classes, the shape of the distribution is barely hinted. An improvement in uniformity will result in an increase in frequency of one or more adjacent classes with an accompanying decrease in the frequency of other classes when multiple classes are used.

The design of the experiment is the same for variable data or attribute data. Any method for selecting the size of the OA, modifying OAs, assigning factors, etc., are utilized for any type of data. The most important differences of the experiments are the sample size required and the analysis of the results.

7-2   Sample Size for Attribute Data

A general rule of thumb for the type of data where occurrences and nonoccurrences are known is that the class with the least frequency should have a count of at least 20. For instance, in an experiment to assess defective parts, the total number of defectives should be at least 20. If the past performance has been 10% defective, then the total sample size of the experiment should be 200 to produce an expected 20 defectives. These defectives are expected to be spread over all the trials to provide information as to which factors were influencing the quantity of defectives. When the attribute data is of the type where only the occurrences may be known, the expected number of occurrences should be from two to five per trial.

In the situation of having a very low failure rate, a very large sample size is required to expect 20 failures in the experiment. For example, if the failure rate during customer use is .25%, then the sample size for the experiment becomes 8000. Rather than use this large a sample size, which may be expensive, it is recommended to intensify the test conditions used in the experimental trials to increase the failure rate and correspondingly decrease the required sample size. The failure mode(s) must remain the same, but the product or process changes that reduce the failure rate on the intensified test should do the same in customer usage. Generally, intensified tests should be used when a reduction in test time and/or sample size is needed (the failure mode must remain the same, which may limit the amount of intensification).

7-3   Analysis of Attribute Data

Attribute data may be analyzed many ways, depending on the type of attribute data (known occurrences and nonoccurrences or known occurrences only) and how many classes are used (two or more than two). The kind of analysis and the kind of data for the situations described are shown in Table 7-1. The most used cases are I and II, case III is used occasionally, and case IV is seldom used. Cases I through III will be discussed in Secs. 7-4 through 7-6, respectively, but case IV will not be discussed.

TABLE 7-1 Attribute Data Analysis Methods

	Known occurrences and nonoccurrences	Known occurrences only

Two classes	Case I
	
	ANOVA
	  0,1 data
	  frequency of occurrence
	  % occurrence
		Case III
		
		ANOVA
		  frequency of occurrence

More than two classes	Case II Accumulation analysis	Case IV Not discussed

7-4   Case Study: Casting Cracks with Two Classes

Two-class attribute data is again the situation corresponding to an all or nothing characteristic, which can be represented numerically by a 1 or 0. In this case the number of occurrences in both classes is known. This case study contains some testing philosophy as well as the Taguchi analytical approach.

7-4-1   Casting problem background

An actual production problem which required the use of attribute data was one concerning cracks in a main structural casting for a heavy-duty automatic transmission. The cracks had not always been a problem but had developed into a 10 to 15% defective castings problem in recent times. Something had definitely changed in the process, but try as they might, the engineers could not resolve the cracking problem. Cracks were sometimes visible with the naked eye, but most were only discernible with magnaflux inspection. Cracks were also more easily noticed after the part had been machined, which made the scrap expense even greater. Recall, from the loss function discussion, that the sooner a problem is discovered, the lower the loss associated with it. Cracks were being found in various locations around the perimeter of the wagon wheel–shaped part; no pattern of the location was apparent.

One hundred percent magnaflux inspection was instituted as a stopgap measure. The expensive magnaflux process was slow and the scrap rate of castings began to interrupt manufacturing and assembly lines. The economics of magnaflux inspection, $2.00 per $17.00 casting, plus the schedule impact, made resolution of this problem very desirable.

7-4-2   Casting cracks experiment structure

A meeting was held to discuss the possible use of a designed experiment to find a permanent solution. The meeting included two casting experts from the manufacturing plant, a foundry representative, a product design engineer, a purchasing agent, and an engineer trained in statistical methods. The focus of the discussion was on what factors might be causing the cracks and the structure of an experiment for investigating these factors. The foundry people were convinced that the use of Taguchi OAs was too complex for the foundry environment and wanted to run simpler two-factor experiments. Other road blocks were thrown up to make their case.

The factors of interest were pouring temperature, time in the mold, cooling rate after removal from the mold, and shot blast intensity. To investigate all these factors two at a time for all possible combinations required six different experiments with 24 different treatment conditions. It was pointed out that an L8 OA could investigate all these things with only eight different treatment conditions. The casting engineers asked how each casting in a trial could be poured at a particular temperature when temperature would fall as castings were poured from the ladle. Data was available which indicated a 40°F temperature drop from the first casting poured to the twelfth and final casting poured. The high and low pour temperatures were revised to indicate the temperature at the start of pouring when the ladle was removed from the holding furnace. If one temperature was better than the other, then this would be the way the foundry would process the castings anyway; one casting at a time would be highly impractical. This procedure introduced some setting error into the experiment but, again, the foundry had to live with this on a day-to-day basis anyway. These were the major roadblocks that were resolved and agreement was made to go ahead with the experiment.

The next step was to select the levels for the casting factors. A summary of the factors and the levels suggested are in Table 7-2. The actual values for the temperature, time, etc., were determined but not recorded here for proprietary reasons.

TABLE 7-2   Casting Cracks Experiment Factors and Levels

Factor		Level 1	Level 2
A	Temperature	Production	Higher
B	Time	Production	Shorter
C	Cooling rate	Production (fan on)	Fan off
D	Shot blast	Production (one cycle)	More cycles

The number of castings needed per trial can be determined by dividing the minimum of 20 defectives for the whole experiment by the past defective rate; 20 divided by .10 portion defective gives a minimum of 200 castings required for the whole experiment. Since 12 castings are obtained per ladle, then 16 ladles would make 192 castings. The 16 ladles would conveniently adapt to an L8 or L16 OA.

7-4-3   Foundry preexperiment

As a historical note, the casting experts in conjunction with the foundry felt that the cracks were caused by pouring the castings too hot, knocking them out of the mold too soon, and cooling them too quickly. To prove their conjecture was correct before the entire experimental effort was "wasted," the foundry poured 20 castings at a high temperature, knocked them out of the mold as soon as possible, and threw them into a snow drift to accelerate cooling. All twenty of the castings were good, entirely crack-free. The experts were quoted as saying "Well, they all didn't crack in production, you know." This was a clue about the casting process but with no other conditions to reference, the understanding is consequently minimal.

7-4-4   Assignment to an orthogonal array

The factors were assigned to an L8 OA with the assignment and experimental results shown in Table 7-3. Recall that two ladles, 24 castings, were poured for each of the trial conditions.

TABLE 7-3   Casting Cracks OA Assignment

	Factors and interactions							No. of castings	
	A	B	$A \times B$	C	e	e	D		
	Column no.							No. of castings	
Trial no.	1	2	3	4	5	6	7	Cracked	Good
1	1	1	1	1	1	1	1	4	20
2	1	1	1	2	2	2	2	1	23
3	1	2	2	1	1	2	2	1	23
4	1	2	2	2	2	1	1	2	22
5	2	1	2	1	2	1	2	4	20
6	2	1	2	2	1	2	1	4	20
7	2	2	1	1	2	2	1	0	24
8	2	2	1	2	1	1	2	0	24
							Totals	16	176

7-4-5   Cursory analysis

In this experiment, trials 7 and 8 provided good results relative to the rest of the experiment and recent production. Note, the common levels for temperature, time, and consequently the interaction of temperature and time. The high pouring temperature in conjunction with short mold time appears to provide good results similar to the "snow drift" experiment and contrary to prior opinions.

7-4-6   Two-class attribute data analysis

A more sophisticated ANOVA for two-class attribute data is as follows. Recall, the bad and good parts can be mathematically represented by 1s and 0s, respectively. Therefore, the data for the first trial is really four 1s and twenty 0s. If 1s and 0s are considered as the data, then the total sum of squares is

$$\mathrm{SS}_T = 1^2 + 1^2 + 1^2 + 1^2 + 0^2 + 0^2 + \cdots + 0^2 - \frac{16^2}{192}$$

Simplified,

$$\mathrm{SS}_T = T - \frac{T^2}{N} = 16 - \frac{16^2}{192} = 14.667$$

The sums of squares for the columns are calculated as with variable data:

$$\mathrm{SS}_A = \frac{(A_1 - A_2)^2}{N} = \frac{(8 - 8)^2}{192} = 0.0$$

$$\mathrm{SS}_B = \frac{(B_1 - B_2)^2}{N} = \frac{(13 - 3)^2}{192} = 0.521$$

$$\mathrm{SS}_{A \times B} = \frac{(A{\times}B_1 - A{\times}B_2)^2}{N} = \frac{(5 - 11)^2}{192} = 0.188$$

$$\mathrm{SS}_C = \frac{(C_1 - C_2)^2}{N} = \frac{(9 - 2)^2}{192} = 0.021$$

$$\mathrm{SS}_D = \frac{(10 - 6)^2}{192} = 0.083$$

There are two sources of error in this experiment which need to be identified. The first source of error, labeled e_1, is from the columns (no main effects assigned but there are interactions) of the OA. The second source, labeled e_2, is from the repetitions in each trial. Intuitively, a form of error can be identified in trial 1, for example. All 24 castings were processed under very similar conditions, yet not all are good or all cracked. If no e_2 were present, then the 24 castings in each trial would be either all good or all bad. Mathematically, the SS_{e_2} amount can be calculated by finding the average of the 0,1 data for each trial, squaring the deviation of the 0,1 data from the average, and summing the squares of the deviations. This is nothing more than an extension of the ANOVA method learned in Chap. 2 (error deviations within groups). In this situation the total variation, SS_T, can be calculated and the variation due to each accountable source, SS_{cols}, subtracted to obtain the variation leftover for repetition error, SS_{e_2}. The degrees of freedom associated with e_2 are determined in a similar manner; the accountable degrees of freedom are subtracted from the total available.

$$SS_{e_1} = \frac{(9 - 7)^2}{192} = 0.021 \qquad \text{Column 5}$$

$$SS_{e_1} = \frac{(10 - 6)^2}{192} = 0.083 \qquad \text{Column 6}$$

A summary of the ANOVA results is shown in Table 7-4. From a practical point of view, the error variance usually doesn't need to be pooled in a classified data analysis. The degrees of freedom associated with e_2 are usually very large and the error variance is changed very little when pooled. Also, the percent contribution may calculate to be a small value, which can be misleading. In this case, if the proper temperature and time combination provide consistently good results, the error under other factor combinations is irrelevant, so care must be taken when

TABLE 7-4   ANOVA Summary of Casting Cracks Experiment

Source	SS	ν	V	F	SS'	p
A*	0.0	1	0.0			
B	0.521	1	0.521	6.95#	0.446	3.05
C*	0.021	1	0.021			
D*	0.083	1	0.083			
$A \times B$	0.188	1	0.188	2.51	0.113	0.78
e_1*	0.104	2	0.052			
e_2*	13.750	184	0.075			
T	14.667	191				100.00
e_p*	13.958	189	0.074			96.17

†At least 90% confidence.
‡At least 95% confidence.
#At least 99% confidence.

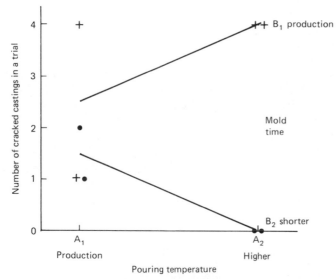

Figure 7-1 Casting cracks experimental plot.

interpreting the percent contribution, which is a function of the average factorial effect minus the averaged error effect.

The ANOVA indicates that time by itself causes a change in defective rate at a high statistical confidence and temperature by itself causes no change in defective rate. The temperature-time interaction is not quite at the 90% confidence level. A plot of the interaction will clarify the interpretation. Note that only one of the factorial combinations is consistent; A_2B_2 has all good parts, the others are some portion good and bad as shown in Fig. 7-1. The error variation averaged over all the treatment conditions masks the contribution of the temperature-time combination. But from a practical point of view, temperature and time actually provide a tremendous contribution toward reducing defective castings when the proper levels are used. This is determined more from the observation method discussed in Sec. 7-4-5 than the ANOVA.

A confirmation experiment was proposed with 300 castings poured at the higher temperature and left in the mold a shorter amount of time. The other two factors, cooling fan and shot blast, were set at the most economical conditions (fan on for faster cooling and subsequent material handling with one pass of shot blast). All 300 castings were defect-free, making a total of 348 castings which were defect free for this process condition. Statistically, this provided a confidence of 97% that the defective rate was less than 1% based on the binomial distribution:

$$P(\text{defects} = 0) = (1 - p)^n = 1 - C$$

where P = percent defective
    n = sample size
    C = confidence level

Because of this mathematical relationship, it is impossible to say at 100% confidence that the defective rate is 0%; only a very large sample size will allow this position to be approached. Even still, this was a substantial improvement over recent performance and may really be 0% defective; just statistically impossible to prove.

An additional 1200 castings were magnafluxed as production parts were poured; all of these were also defect-free (now 97.9% confidence that the percent defective is less than .25%). At this point magnaflux was discontinued and the casting process allowed to perform at the 100% yield conditions with respect to casting cracks.

7-4-7   Casting experiment summary

In retrospect, the snow drift experiment that had been run earlier is understandable. The conditions in the snow drift experiment for pour temperature and mold time were very similar to conditions in trials 7 and 8 for the L8 OA. The OA experiment, however, allowed the relationship between temperature and time to be comprehended. The casting experts were so intent on a preconceived notion of which factors and conditions were causing a problem that they were preventing themselves from deriving a solution. In fact, the production conditions for temperature and time suggested by the engineers who were led by their experience and intuition caused 10.4% (5/48) of the castings to crack in the experiment. This was very similar to the recent production history. One advantage of an OA experiment is its objectivity. Some combinations of factors and levels are tested which otherwise may not have been investigated. Evidently, the foundry had lost control of temperature and/or time, which caused some bad castings to show up. When the casting experts took over and used experience of other casting problems to adjust the process, the defect problem actually became worse.

Another aspect of these experimental results is that the casting experts could not explain why the pouring conditions of higher temperature and shorter time prevented cracks. The knowledge of why cracks are caused or why they can be prevented is irrelevant if what to do to prevent cracks is known. Even the fact that an interaction between temperature and time may exist becomes irrelevant. Knowing what temperature and time consistently prevents cracks is the practical information that can be put to use on the factory floor. Of course, if the reason why can be determined, then this information can be used in other casting designs and development. This critique of the casting experts is not meant to belittle their knowledge or experience; one must under-

TABLE 7-5   Crack Severity Rating

Severity	Class no.
No cracks	1
Mild crack(s)	2
Moderate crack(s)	3
Severe crack(s)	4

stand the physics of a design or process to know which factors to investigate. Understanding the problem areas will, hopefully, encourage the use of more effective experimentation methods when faced with a problem.

7-4-8   Attribute data transformation

Case I mentions transformation of the 0,1 data to other forms of data which become variable and can be analyzed by the typical ANOVA methods of Chap. 2. The actual frequency of occurrences in a trial is a discrete variable, the percent of occurrences in a trial is a continuous variable within the limits of 0.0 to 100.0%.

7-5   Castings Cracks with Multiple Classes

The data for the casting experiment was also collected in a fashion which described the severity of the cracks that appeared in the casting. Accumulation analysis takes frequency and severity of occurrences into account and is used when the class number has an engineering meaning. The classes for crack severity are described in Table 7-5 along with the accompanying rating value. The frequency of the castings falling into these various categories is then tallied in Table 7-6. Trial 6 resulted in 20 castings with no cracks, 2 castings with mild cracks, and 2 castings with moderate cracks. The other trials are interpreted accordingly.

The first step in accumulation analysis is to create a cumulative frequency table by summing the frequencies left to right (or right to left) for each trial. The cumulative values are calculated:

$$\text{Class I} = \text{class 1}$$

$$\text{Class II} = \text{class 1} + \text{class 2}$$

$$\text{Class III} = \text{class 1} + \text{class 2} + \text{class 3}$$

$$\text{Class IV} = \text{class 1} + \text{class 2} + \text{class 3} + \text{class 4}$$

Table 7-7 shows the cumulative frequencies for the classes which were determined by summing from left to right. One will notice that the last

TABLE 7-6 Frequency of Occurrence of Castings

		Class			
Trial no.	1	2	3	4	Totals
1	20	2	1	1	24
2	23	0	0	1	24
3	23	0	1	0	24
4	22	0	2	0	24
5	20	4	0	0	24
6	20	2	2	0	24
7	24	0	0	0	24
8	24	0	0	0	24
Totals	176	8	6	2	192

class shows no variation from trial to trial. For this reason, in accumulation analysis the number of classes minus one is analyzed.

The second step is to calculate a weight value for each class, which is a function of the cumulative frequency of occurrence in that class. The formula for the weight for class i is

$$W_i = \frac{N^2}{T_i(N - T_i)}$$

where N = total number of tests
  T_i = accumulative occurrences in class i

Therefore,

$$W_{\mathrm{I}} = \frac{N^2}{T_{\mathrm{I}}(N - T_{\mathrm{I}})} = \frac{192^2}{176(192 - 176)} = 13.09$$

TABLE 7-7 Cumulative Frequency of Occurrence of Castings

	Cumulative class			
Trial no.	I	II	III	IV
1	20	22	23	24
2	23	23	23	24
3	23	23	24	24
4	22	22	24	24
5	20	24	24	24
6	20	22	24	24
7	24	24	24	24
8	24	24	24	24
Totals	176	184	190	192

$$W_{\text{II}} = \frac{N^2}{T_{\text{II}}(N - T_{\text{II}})} = \frac{192^2}{184\,(192 - 184)} = 25.04$$

$$W_{\text{III}} = \frac{N^2}{T_{\text{III}}(N - T_{\text{III}})} = \frac{192^2}{190\,(192 - 190)} = 97.01$$

The sum of squares for each class is multiplied by the class weight to obtain a weighted sum of squares for each class. The weighted sums of squares are then added to obtain the complete sum of squares.

The third step is to calculate the total sum of squares:

$$SS_T = \sum_{i=1}^{nca} SS_{T_i} W_i$$

where nca is the number of classes analyzed. Similar to two-class analysis,

$$SS_{T_i} = T_i - \frac{T_i^2}{N}$$

Then in this example,

$$SS_T = SS_{T_{\text{I}}} W_{\text{I}} + SS_{T_{\text{II}}} W_{\text{II}} + SS_{T_{\text{III}}} W_{\text{III}}$$

Substituting, but showing the detail in the first class only,

$$SS_T = \left(T_{\text{I}} - \frac{T_{\text{I}}^2}{N}\right)\left(\frac{N^2}{T_{\text{I}}(N - T_{\text{I}})}\right) + (\quad)(\quad) + (\quad)(\quad)$$

$$= T_{\text{I}}\left(1 - \frac{T_{\text{I}}}{N}\right)\left(\frac{N^2}{T_{\text{I}}(N - T_{\text{I}})}\right) + (\quad)(\quad) + (\quad)(\quad)$$

$$= T_{\text{I}}\left(\frac{N - T_{\text{I}}}{N}\right)\left(\frac{N^2}{T_{\text{I}}(N - T_{\text{I}})}\right) + (\quad)(\quad) + (\quad)(\quad)$$

$$= N + N + N$$

$$= N(nca) = 192(3) = 576.0$$

The fourth step is to calculate the OA column sums of squares. The sums of squares for each class are computed the same as two-class attribute data. The column sums of squares are multiplied by the class weight and added together to get the complete sums of squares:

$$SS_{A_i} = \frac{(A_{1_i} - A_{2_i})^2}{N}$$

where A_{1_i} = accumulative occurrences for the first level of factor A in class i.

$$SS_A = \sum_{i=1}^{nca} SS_{A_i} W_i$$

Referring to the original factorial assignment of the OA,

$$SS_{A_I} = \frac{(88 - 88)^2}{192} = 0.0$$

$$SS_{A_{II}} = \frac{(90 - 94)^2}{192} = 0.0833$$

$$SS_{A_{III}} = \frac{(94 - 96)^2}{192} = 0.0208$$

$$SS_A = 0.0\,(13.09) + 0.0833\,(25.04) + 0.0208\,(97.01)$$

$$= 4.1000$$

The remaining columns are calculated in a similar fashion.

The degrees of freedom in multiple-class analysis are similar to the other methods. However, each attribute class allows a comparison of the first and second levels of each OA column, which means each class analyzed provides a degree of freedom. Therefore, each column in the array has a similar calculation to factor A

$$\nu_A' = \nu_A\,(nca) = 1\,(3) = 3$$

The total degrees of freedom is

$$\nu_T' = \nu_T\,(nca) = (N - 1)\,(nca) = 191\,(3) = 573$$

The ANOVA summary is shown in Table 7-8. In this instance, the conclusions from the accumulation analysis are nearly the same as the two-class analysis. However, as demonstrated by this case study, when the frequency of occurrence is predominantly one of the end classes, the accumulation analysis doesn't offer any advantage over a simpler two-class analysis. Accumulation analysis works best when the distribution of frequency of occurrence is spread over many classes.

7-6   Two-Class Attribute Data; Known Occurrences Only

In this situation the frequency of occurrences can be treated the same as variable data utilizing a standard ANOVA. Examples of this situation would be the number of surface imperfections per part for a material handling system or possibly the number of piglets produced by a sow at farrowing. The last situation would be a very interesting experiment!

TABLE 7-8 ANOVA Summary, Multiple Class Attribute Data

Source	SS	v	V	F
A	4.100	3	1.370	1.370
B	9.360	3	3.120	3.120‡
C	0.790	3	0.260	
D	3.180	3	1.060	
$A \times B$	5.000	3	1.667	1.667
e_1	1.880	6	0.310	
e_2*	551.690	552	1.000	
T	576.000	573		

*e_2 used for F test.
†At least 90% confidence.
‡At least 95% confidence.
#At least 99% confidence.

7-7 Summary

This chapter covered the typical situations encountered in using attribute date. The classified data experiment is structured identically to variable data experiments. The analysis of the results is the fundamental difference between attribute and variable data experiments. The next chapter introduces the most powerful aspect of the Taguchi methodology of product and process development.

Problems

7-1 If factors A to G are assigned to an L8 OA and the data for each trial is (two classes)

Trial no.	1	2	3	4	5	6	7	8
Class 1	3	4	3	2	7	8	7	9
Class 2	7	6	7	8	3	2	3	1

What would be the ANOVA results?

7-2 If factors A to G are assigned to an L8 OA and the data for each trial is (four classes)

Trial no.	1	2	3	4	5	6	7	8
Class 1	2	1	2	2	4	4	4	5
Class 2		1	3	2			1	
Class 3	2	2		1		1		
Class 4	1	1			1			

What would be the ANOVA results?

Parameter and Tolerance Design

8-1 Parameter and Tolerance Design
Explanation

This chapter concerns probably the largest contribution to quality methodology that Taguchi has made. The designed experiment addressed in all of the previous chapters but the first has existed since the 1930s. However, previous experimental approaches looked upon all factors as causes of variation. If these causes could be well controlled or eliminated, then product or process variation would be reduced, and, therefore, quality would be improved. But if a product is sensitive to ambient temperature variations, how can anyone control or eliminate temperature in a customer's environment? The answer is obvious; ambient temperature variations can neither be controlled nor eliminated without a large expense. Ambient temperature in a factory might be controlled very well to eliminate the temperature effect on a machine's performance, but the world's atmosphere cannot be controlled, so all kinds of devices are exposed to large temperature variations. Therefore, a different approach is required if product quality is to be improved. This approach was entitled parameter design by Taguchi and is covered in the majority of this chapter. Parameter design is used to improve quality without controlling or eliminating causes of variation. Controlling or eliminating causes of variation may be expensive compared to a parameter design approach.

Taguchi views the design of a product or process as a three-phase program:

1. System design
2. Parameter design
3. Tolerance design

System design is the phase when new concepts, ideas, methods, etc., are generated to provide new or improved products to customers. One way to remain competitive in the world economy is to be a leader in utilizing technology. However, the technological advantage disappears quickly because it may be copied. If a competitor can fabricate the same new idea in a more uniform manner, then the technological advantage is more than lost. The parameter design phase is crucial to improving the uniformity of a product and can be done at no cost or even at a savings. This means certain parameters of a product or process design are set to make the performance less sensitive to causes of variation. The tolerance design phase improves quality at a minimal cost. Quality is improved by tightening tolerances on product or process parameters to reduce the performance variation. This is done only after parameter design.

Typically, when a problem is detected in product development, an engineer may jump directly to tolerance design; when tolerances are tightened, variation will be reduced and quality improved. However, tightening tolerances may be expensive and completely unnecessary if parameter design were used first. One serious mistake a designer can make is to use expensive materials, components, or processes for a product when lower-cost items may be used if a parameter design approach is applied.

8-2 Control and Noise Factors

To begin the discussion on parameter design, one must consider kinds of design and development factors. Taguchi separates factors into two main groups: control factors and noise factors. Control factors are those which are set by the manufacturer and cannot be directly changed by the customer. An engine manufacturer can dictate the material for the pistons, the tension on the piston rings, the piston-bore clearance, etc., which cannot be easily modified by the customer (of course, many hot-rod fanatics do this very thing to improve short-term output horsepower usually at the sacrifice of durability). Noise factors are those over which the manufacturer has no direct control but which vary with the customer's environment and usage. In general, noise factors are those which the manufacturer desires not to have to control at all.

Noise factors can be placed in three categories:

1. Outer noise
2. Inner noise
3. Product noise

Outer noises are environmental factors such as ambient temperature, humidity, pressure, or people. Even different batches of materials may be viewed as outer noise to a production process. Inner noises are function- and time-related, such as deterioration, wear, fade of color, shrinkage, and drying out. Outer noises produce variation from outside the product; inner noises produce variation from inside or within the product; and product noise manifests itself in part-to-part variation. Products may have sensitivity to all three forms of noise simultaneously. The design quality of a product or process provides less functional variation due to outer or inner noise. Production quality provides less functional variation from one part to another and close to a target value. Taguchi refers to design quality efforts as off-line quality control (QC) and production quality efforts as on-line quality control. Off-line QC is any of the design and development activities that may take place before products are manufactured and available to customers. When products are manufactured to be available for customers, the on-line activities begin. Of course, the on-line activity must be well planned before the start of production. The loss function reflects the off- and on-line QC efforts in the equation:

$$L(y) = k[S_y^2 + (\bar{y} - m)^2]$$

The variance S_y^2 is reduced and the target value m determined during the off-line phase. The average value produced $\bar{y}$ is controlled during the on-line phase. Both of these phases will reduce the loss associated with the product. The real purpose of a quality control chart is to provide the proper, close to the target, average value from a process. Incidental (special) causes of variation may be identified and controlled or eliminated, but this is the tolerance design approach. If a large amount of variation is present at the introduction of a process, this is an indication of the lack of off-line QC efforts. The more off-line QC work done, the more robust a process or product is against disturbances (outer and inner noise) in the environment and life of a product. Again, to reduce loss to society in the loss function there must be reduced variance (off-line QC work) and an average near the target value (on-line QC work).

Parameter and tolerance design take on additional meanings with the loss function approach. Parameter design is used to dampen the effect of noise (reduce variance) by choosing the proper level for control factors. Parameter design is used to improve quality without controlling or removing the cause of variation, to make the product robust against noise factors. Tolerance design is to reduce the variation by tightening tolerances around the chosen target value of the control factors. Tolerance design reduces or eliminates the effect of causes of variation. By using parameter and tolerance design, the true critical characteristics (control factors) can be identified and minimized in number.

8-3 Introduction to Parameter Design

Imagine that an experiment is structured like that shown in Table 8-1 with control factors only assigned to the L8 OA. Why bother to assign noise factors when they cannot be controlled or eliminated without some expense? Three repetitions for each trial are planned in this instance. What will happen as data points y_1, y_2, and y_3 are collected for each trial (not necessarily one after another; remember the randomization strategies)? Will y_1 be identical to y_2 and identical to y_3? Noise effects, all those uncontrolled and unknown factors that were actually varying from one degree to another as the experiment was run, cause the differences from data point to data point. Rather than assuming that error variation is an aggregate of all these noise effects and equally distributed in all treatment conditions, parameter design utilizes these repetitions to aid in identifying what levels of what control factors might have reduced variation. There may be some control factor(s) of which one of the levels is more robust against the noise effects.

In the previous example repetitions were used to assess the noise effect on some performance characteristic(s) of interest. However, these noise effects were merely accidental as the experiment was being con-

TABLE 8-1 Simple Parameter Design Experiment

	Control factors							Data		
	A	B	C	D	E	F	G			
	Column no.									
Trial no.	1	2	3	4	5	6	7	y_1	y_2	y_3
1	1	1	1	1	1	1	1	*	*	*
2	1	1	1	2	2	2	2	*	*	*
3	1	2	2	1	1	2	2	*	*	*
4	1	2	2	2	2	1	1	*	*	*
5	2	1	2	1	2	1	2	*	*	*
6	2	1	2	2	1	2	1	*	*	*
7	2	2	1	1	2	2	1	*	*	*
8	2	2	1	2	1	1	2	*	*	*

*Data points.

TABLE 8-2   Inner/Outer OA Parameter Design Experiment

								L4 OA outer array (noise factors)			
							Z	1	2	2	1
							Y	1	2	1	2
							X	1	1	2	2
	L8 OA inner array (control factors)										
	A	B	C	D	E	F	G				
	Column no.							Data			
Trial no.	1	2	3	4	5	6	7	y_1	y_2	y_3	y_4
1	1	1	1	1	1	1	1	*	*	*	*
2	1	1	1	2	2	2	2	*	*	*	*
3	1	2	2	1	1	2	2	*	*	*	*
4	1	2	2	2	2	1	1	*	*	*	*
5	2	1	2	1	2	1	2	*	*	*	*
6	2	1	2	2	1	2	1	*	*	*	*
7	2	2	1	1	2	2	1	*	*	*	*
8	2	2	1	2	1	1	2	*	*	*	*

ducted; error variance is all the unknown and uncontrolled factors which are varying in unknown amounts. A better parameter design strategy would be to force noise effects into the experiment in a different manner. Table 8-2 shows an experimental layout with an inner array for control factors only and an outer array for noise factors only. If these factors are all mixed in an inner array, then this is traditional cause detection experiment. If a product is found to be sensitive to ambient temperature, what then? Can instructions be supplied with the product to tell the customer only to use the product within a certain temperature range? Not many people would buy such a product.

This parameter design strategy separates the control factors from the noise factors by using inner and outer arrays, respectively. Now noise factors (X, Y, and Z) such as temperature and operators could be assigned to the outer array to find some level of a control factor that doesn't have much variation in the results in spite of the noise factors definitely being present.

In this experimental arrangement there are 32 separate test conditions. A trial number specifies a test condition with respect to control factors, but the outer array specifies four test conditions with respect to noise factors for that trial. If tests are very expensive to run, a thorough outer array may be precluded and only one noise factor thought to be strong (or noise factors combined into the best and worst conditions) may be used.

Also, if product or process performance can be modeled mathematically, then a much less expensive experiment can be run. The equation, i.e., mathematical model, of a system will have many terms with various powers and magnitudes of coefficients for those terms. The designed experiment can use the terms as factors and substitute into the equation high and low values as prescribed by the OA. Obviously, repetitions are not necessary since the equation will give exactly the same result each time. The performance of electronic devices, voltage, amperage, etc., can be readily modeled mathematically, which makes the electronics field a prime candidate for designed experiments using simulation. In addition, electronic devices abound with nonlinear performance characteristics, making a parameter design approach very desirable. The nonlinear aspect of parameter design will be discussed later in Sec. 8-6-1. Hardware should be used on a few of the test conditions to verify the mathematical model.

8-4   Signal-to-Noise Ratios

The control factors that may contribute to reduced variation (improved quality) can be quickly identified by looking at the amount of variation present as a response. All past analyses in this text have addressed which factors might affect the average response, but now there is interest in the effect on variation as well. Taguchi has created a transformation of the repetition data to another value which is a measure of the variation present. The transformation is the signal-to-noise (S/N) ratio. The S/N ratio consolidates several repetitions (at least two data points are required) into one value which reflects the amount of variation present. There are several S/N ratios available depending on the type of characteristic; lower is better (LB), nominal is best (NB), or higher is better (HB).

8-4-1   Different S/N Ratios

The S/N ratio, which condenses the multiple data points within a trial, depends on the type of characteristic being evaluated. The equations for calculating S/N ratios for LB, NB, or HB characteristics are:

1. Lower is better

$$S/N_{\mathrm{LB}} = -10 \log \left(\frac{1}{r} \sum_{i=1}^{r} y_i^2 \right)$$

where r = number of tests in a trial (number of repetitions regardless of noise levels.

2. Nominal is best

$$S/N_{\mathrm{NB_1}} = -10 \log V_e \qquad \text{(variance only)}$$

$$S/N_{\mathrm{NB_2}} = +10 \log \left(\frac{V_m - V_e}{rV_e} \right) \qquad \text{(mean and variance)}$$

3. Higher is better

$$S/N_{HB} = -10 \log \left(\frac{1}{r} \sum_{i=1}^{r} \frac{1}{y_i^2} \right)$$

The LB and HB S/N ratios are both easy to calculate; each repetition is entered into the equation. However, the NB S/N ratio needs further explanation. Both ratios contain the value V_e, and NB_2 contains V_m. These values are determined by doing a no-way ANOVA (see Sec. 2-2) on all the repetitions for a trial. Even though the no-way ANOVA method appeared to have no real purpose in the analysis discussion, the S/N ratio utilizes this decomposition approach. Recall that the variance due to the mean V_m has 1 degree of freedom associated with it always; therefore,

$$V_m = \frac{SS_m}{V_m} = \frac{S_m}{1} = SS_m = r(\bar{y})^2$$

$$V_e = \frac{SS_e}{V_e} = \left(\frac{SS_T - SS_m}{r - 1} \right)$$

$$SS_T = \sum_{i=1}^{r} y_i^2$$

$$S/N_{NB_1} = -10 \log \left(\frac{SS_T - SS_m}{r - 1} \right)$$

$$S/N_{NB_2} = +10 \log \left(\frac{SS_m - V_e}{rV_e} \right)$$

The S/N for NB_1 is a function of variation only and the S/N for NB_2 is a function of both average and variation. For this reason, the use of S/N_{NB_1} is recommended by Hunter for the NB type of characteristic.* The log V_e also provides a tendency toward normality for the variance statistic. The S/N_{NB_1} might be considered as a substitute for LB or HB characteristics since log V_e is independent of the average but S/N_{LB} and S/N_{HB} are not. The following example demonstrates the effect of using the two different S/N_{NB} ratios.

8-4-2 Example *S/N* calculation

Four possible trial results are shown which have differing amounts of variation in an experiment arranged with an inner array and a two-level noise factor array, N_1 and N_2.

---

*J. Stuart Hunter, "Signal-to-Noise Ratio Debated," *Quality Progress*, May 1987, pp. 7–9.

Trial 1 results:

N_1 11 10 9 N_2 11 10 9

$$SS_T = \sum_{i=1}^{r} y_i^2 = 604$$

$$SS_m = r(\bar{y})^2 = 6(10)^2 = 600$$

$$V_e = \left(\frac{SS_T - SS_m}{r-1}\right) = \frac{(604 - 600)}{5} = .8$$

$$S/N_{NB_1} = -10 \log V_e = -10 \log .8 = .97 \text{ db}$$

$$S/N_{NB_2} = +10 \log \left(\frac{SS_m - V_e}{rV_e}\right) = 10 \log \left[\frac{600 - .8}{6(.8)}\right] = 20.96 \text{ db}$$

Trial 2 results:

$$N_1 \ 12 \ 10 \ 8 \qquad N_2 \ 12 \ 10 \ 8$$

$$S/N_{NB_1} = -5.05 \text{ db}$$

$$S/N_{NB_2} = 14.93 \text{ db}$$

Note that both S/N ratios dropped virtually the same amount of dbs due to the increase in variability; the average remained exactly the same. As variation is doubled, the S/N ratio drops approximately 3 db.

Trial 3 results:

$$N_1 \ 12 \ 11 \ 10 \qquad N_2 \ 10 \ 9 \ 8$$

$$S/N_{NB_1} = -3.02 \text{ db}$$

$$S/N_{NB_2} = 16.98 \text{ db}$$

Note that the S/N ratios are not as high as trial 1, but are higher than trial 2. The average value is exactly the same, but variation is more than trial 1 and less than trial 2.

Trial 4 results:

$$N_1 \ 21 \ 20 \ 19 \qquad N_2 \ 21 \ 20 \ 19$$

$$S/N_{NB_1} = .97 \text{ db}$$

$$S/N_{NB_2} = 26.99 \text{ db}$$

In this case, the variation is the same as trial 1 but the average has increased. One can see that S/N_{NB_1} is a function of the variability only and S/N_{NB_2} is a function of both variability and average. Table 8-3 summarizes the results of this example in which it has been demonstrated that reduced variation relative to the average causes an increase in the S/N value; S/N is always a HB characteristic (not the absolute value in the case of negative db values; -3.02 db is higher than -5.05 db).

8-4-3 Analysis of *S/N* ratio as a response

The S/N ratio is treated as a response of the experiment, which is a measure of the variation within a trial when noise factors are present. If an outer array is used, the noise variation is forced in an experiment; with pure repetitions (no outer array), the noise variation is unforced. S/N ratio is a response which consolidates repetitions and the effect of noise levels into one data point. A standard ANOVA can be done on the S/N ratio which will identify factors significant to increasing the average value of S/N and subsequently reducing variation.

8-5 Parameter Design Strategy

When the ANOVA on the raw data (identifies control factors which effect average) and the S/N data (identifies control factors which effect variation) are completed, the control factors may be put into four classes:

Class I: Factors which affect both average and variation

Class II: Factors which affect variation only

Class III: Factors which affect average only

Class IV: Factors which affect nothing

The parameter design strategy is to select the proper levels of class I and II to reduce variation and class III to adjust the average to the target value. Class IV may be set at the most economical level since nothing is affected. Figure 8-1 shows example plots for the four classes of factors. The fundamental parameter design approach is to move all but one design (control) parameter to a region of low response slope to make

TABLE 8-3 Nominal-Is-Best *S/N* Ratios

Trial no.	Data		*S/N*	
	N_1	N_2	NB_1	NB_2
1	11, 10, 9	11, 10, 9	0.97	20.96
2	12, 10, 8	12, 10, 8	-5.05	14.93
3	12, 11, 10	10, 9, 8	-3.02	16.98
4	21, 20, 19	21, 20, 19	0.97	26.99

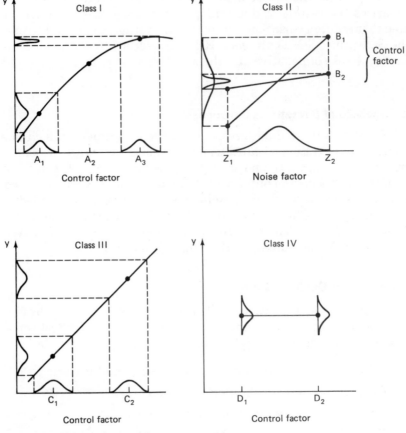

Figure 8-1 Classes of control factors.

those parameters insensitive to variation. The remaining factor(s) should be linear (adjustment) factors to obtain the appropriate average response (quality characteristic). When starting with two-level experiments, insignificant factors are already parameter designed and significant factors may be investigated further to discover nonlinear properties, if possible.

8-6 Case Studies of Parameter Design

Two case studies to be discussed use the property of the class I control factor primarily to reduce variation. Both of these case studies require the use of a class III factor to adjust the average, but the case study on heat treatment is an interesting application. The third case study utilizes a class II factor for parameter design.

8-6-1   Case study: electronic circuit*

A power supply circuit is required to provide a certain output voltage to another electronic circuit. A designed experiment has been done on all the electronic components to determine some significant factors in the various classes. This experiment could have been done by modeling the circuit mathematically and calculating the output voltage when components take on different levels rather than testing physical hardware. Again, electronic circuitry is particularly adaptable to this approach and a mathematical model should be used whenever possible to reduce test costs and time.

Imagine that two components are strongly related to output voltage performance as determined by a designed experiment. The voltage could be a function of transistor gain, component A, which was evaluated at three levels of gain and resistance, component B, which was evaluated at two levels as shown in Fig. 8-2. The curves labeled B_1 and B_2 are generated from the average voltages obtained under the six possible treatment conditions of factors A and B. Based on this performance, a typical designer might specify a transistor gain of A_x to provide the targeted average voltage. However, since transistors vary in gain from one part to another, the output voltage will vary from assembly to assembly due to transistor gain variation as indicated by distribution I.

A parameter design approach, in contrast, would be to specify the transistor gain in the neighborhood of A_z. At this value of gain the variation of transistors could be even greater, but the variation of output voltage would be substantially reduced as indicated by distribution II. The problem here is that the average voltage produced is greater than the target value, so an "adjustment" factor is required. Resistance may be altered to the value B_w to reduce the average voltage to the target value. In this case the transistor gain is a class I factor and resistance is a class III factor.

Several advantages are gained by the parameter design approach. First, larger variation of transistor gain is not passed on to the variation in output voltage. This allows lower-quality components to be used without lowering the quality of the overall circuit. This makes the circuit robust against transistor product noise. Making a circuit robust against one noise makes it robust against other noises also. If transistor gain changes slightly with operating temperature, an outer noise, then this will not be passed on to variation in output voltage. If transistor gain changes slightly with age, an inner noise, this will not be passed on either. Second, because the sensitivity of transistor gain is greatly reduced, the number of critical components in the circuit is reduced. Using

---

*Genichi Taguchi and Yu-in Wu, *Off-line Quality Control*. Central Japan Quality Control Association, Nagaya, 1979, pp. 30–31.

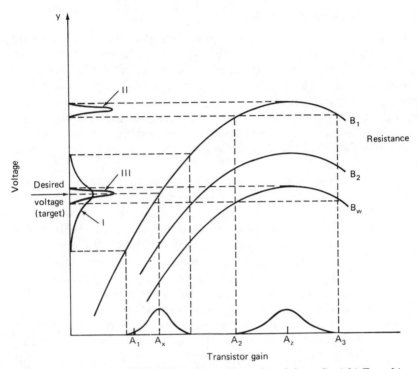

Figure 8-2 Electronic circuit performance. (*Reproduced from Genichi Taguchi and Yu-in Wu,* Off-Line Quality Control. *Central Japan Quality Control Association, Nagaya, 1979, p. 30. Used by permission.*)

the parameter design approach minimizes the number of critical factors in a product or process, which subsequently reduces the cost of the on-line QC efforts.

At this point, tolerance design should be applied if variation due to transistor gain and/or resistance causes excessive variation in output voltage. Tolerance design is simply reducing the variation in a component to reduce variation in the system. The improved quality required of some of the components will increase the cost of the system. The loss function should be used to justify any cost increases due to higher-quality components.

Now it should be apparent why parameter design must be applied before tolerance design. If the original level of A_x were used as the target value for transistor gain and tolerance design applied first, then the component costs of both the transistor and resistor would increase. The tolerances on the transistor and resistor would have to be tightened to reduce variation in output voltage. However, now the circuit is still sensitive to outer and inner noise because of utilizing the high slope area of the performance curve.

Parameter design should be applied first to improve quality at no additional cost. Tolerance design should be applied second, if necessary, to improve quality at a minimal cost. The nonlinear performance characteristic is one that should be utilized to the fullest extent for low-cost quality improvement.

8-6-2 Case study: heat treatment

This case study involved the use of a nonlinear characteristic of a process. A heat treatment process, carbonitriding, was primarily used to create wear-resistant surfaces on an engine component. Unfortunately, a side effect of the heat treatment process caused a slight growth in height of the component. The predictability of this growth had been causing a scrap problem after the heat treatment operation. An experiment was designed to investigate various heat treatment factors.

Out of all those studied, the testers screened out one key factor which had a nonlinear tendency with respect to change in height (growth) during heat treatment. Three levels of the amount of ammonia (NH_3) introduced into the heat treatment atmosphere during a batch were evaluated with results shown in Fig. 8-3. The past production level had been at the 5 ft³ (.1415 m³) value. The device which measured the amount of NH_3 introduced into the heat treatment atmosphere had a precision of approximately ±1 ft³ (.0283 m³). The variation in amount of ammonia caused substantial variation in batch-to-batch change in height.

The parameter design approach was utilized and a level of 8.75 ft³ (.2476 m³) was chosen for the production conditions. Any value above 10 ft³ (.2830 m³) caused adverse chemical reactions. At the new level the measuring device doesn't have to be replaced or improved to cause an improvement in the quality of the heat treatment process. Only the average change in height had to be accommodated. The adjustment factor was easy to come by; only the pre-heat treatment height had to be changed to handle the increased amount of growth. The parts would have to be machined to a slightly smaller value than in the past, but the resultant part after heat treatment would be much more consistent batch to batch.

Increasing the amount of NH_3 per batch did increase costs; NH_3 costs approximately $.0025 per cubic foot ($.0883 per cubic meter). For approximately one cent per batch, the quality of this process was vastly improved. The tolerance design approach of improving the NH_3 measuring device would have entailed a much greater cost than one cent per batch. The subsequently reduced inspection costs and scrap rate saved approximately $20,000 annually for the amount of one cent per batch investment.

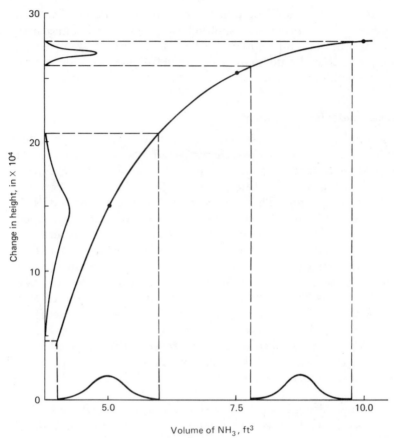

Figure 8-3 Ammonia content versus growth.

8-6-3 Case study: fishing reel line roller

Two fishing reels of identical design but different sizes were purchased from the same company. The reels were open-faced spinning reels of intermediate and large line capacity and operated identically with respect to the line retrieval mechanism. Line is wound around a stationary spool (which has an axis parallel to the fishing pole) by running the line over a roller that revolves around the spool on a bail mechanism. The roller axis of rotation is nominally perpendicular to the line spool and fishing pole axes but does not intersect the axis of the line spool. Figure 8-4 shows two views of the spool and roller mechanism for line retrieval. The roller axis is nominally perpendicular to the spool axis, but one reel assembly by chance had a positive roller angle and the other by chance a negative roller angle. The intent is to have the line run in the center of the hourglass-shaped roller to prevent line deterioration through many

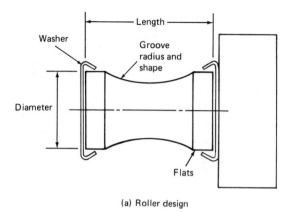

(a) Roller design

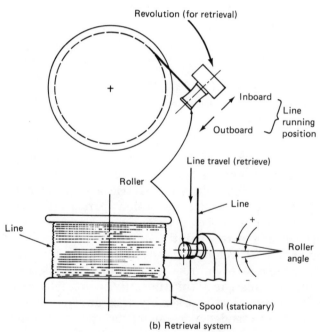

(b) Retrieval system

Figure 8-4   Fishing reel design. (*a*) Roller design; (*b*) retrieval system.

retrieves. However, the roller with the positive angle would tend to allow the line to run on the outboard end against the washer and the roller with the negative angle would tend to allow the line to run on the inboard end against the washer. A plot of this performance is shown in Fig. 8-5 as the current roller design. The line running against the side washers would deteriorate very rapidly and perhaps under high load

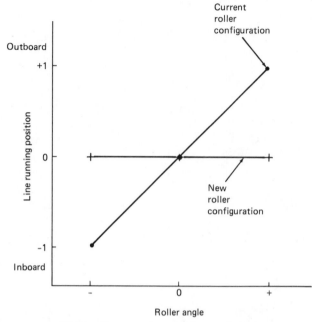

Figure 8-5   Fishing line running position performance.

(perish the thought of it being a large steelhead trout) would separate. The reels were reworked, shimming and filing, by the customer to achieve a zero roller angle, which provided a line retrieval position in the center of the roller. This again was a tolerance design approach (tightening tolerances to reduce performance variation), but it was the only alternative the customer had at this point.

A parameter design approach would treat the roller angle as a product noise factor (desiring not to have to control the roller angle). A tolerance design approach would be a mistake at this time; tightening tolerances on roller angle by adjustment, selective assembly, or scrap of excessively angled parts to eliminate the tendency to run off the end of the roller

TABLE 8-4   **Fishing Reel Parameter Design Factors and Levels**

Roller factors		Level 1	Level 2
A	End flats	Yes	No
B	Groove radius	.250 in (6.35 mm)	.375 in (9.53 mm)
C	Groove shape	Radius	Vee
D	Diameter	.250 in	.375 in
E	Length	.375 in	.500 in (12.70 mm)

would add unnecessary cost. A parameter design approach would be to find a roller configuration that would be insensitive to roller angle with regard to line running position. A new high-quality roller configuration, as yet undetermined, is also shown in Fig. 8-5. The line would run in the center in spite of roller angle variation. This shows the class II type of control factor which interacts with a noise factor. In this case, the roller configuration factor(s) would be the control factor(s) and roller angle would be the noise factor. A recommended experiment to determine the appropriate roller configuration might include the factors shown in Table 8-4 and the experimental layout might look like the OA in Table 8-5.

In this proposed experiment the inner array is a typical L8 OA and the outer array is a one-factor, three-level array. To conduct the experiment, the various roller configurations would have to be manufactured, the line retrieved under load, with the roller angle adjusted to the $-$, 0, and $+$ positions, and the running position of the line observed. A response value of $+1$ could be assigned if the line ran off the outboard end, a value of 0 assigned if the line ran in the center, and a -1 assigned if the line ran off the inboard end. The data could then be analyzed by ANOVA of the S/N value for a NB characteristic.

This experimental approach would reveal if any control factor(s) would provide a roller design that would be more robust against roller angle variation without necessarily increasing cost. An angler using such a roller design would notice that the reel is robust against other noises as well. Other noises are line diameter (strength), speed of retrieve, load on the line, amount of line currently on the spool, etc. This approach to reel design would provide a much higher quality reel at no cost.

TABLE 8-5 Fishing Reel Parameter Design Experiment

	Roller factors							Roller angle	
	A	B	C	D	E	e	e			
	Column no.								Roller angle	
Trial no.	1	2	3	4	5	6	7	$-$	0	$+$
1	1	1	1	1	1	1	1	*	*	*
2	1	1	1	2	2	2	2	*	*	*
3	1	2	2	1	1	2	2	*	*	*
4	1	2	2	2	2	1	1	*	*	*
5	2	1	2	1	2	1	2	*	*	*
6	2	1	2	2	1	2	1	*	*	*
7	2	2	1	1	2	2	1	*	*	*
8	2	2	1	2	1	1	2	*	*	*

*Data points.

8-7   Analysis of Inner/Outer Array Experiment

The ANOVA for an inner/outer OA experiment is somewhat more complex than an inner OA only because of the additional sources of variation. The outer array factors add sources of variation, but because of the structure of the inner/outer array, there are many additional two-factor interactions. Each inner array factor interaction with an outer array factor may be estimated; the inner/outer array structure is a full factorial experiment for control and noise factors.

8-7-1   Case study: die cast component

The experiment referenced in Sec. 2-4-1 was only a portion of an actual experiment which had the objective of increasing the Rockwell hardness (B scale) of a die cast engine component. A total of five factors and one interaction were of interest, as well as the consistency of the hardness from position to position on the part. The five factors and levels chosen are listed in Table 8-6. The copper-magnesium interaction was of interest and also two critical areas (positions) on the casting. The levels for the metallic factors were the Society of Automotive Engineers' (SAE) specification limits for the specified alloy. The levels for water and air cooling were process-dependent. The experimental layout and the results for three repetitions in each treatment condition are shown in Table 8-7. Note that only three pistons were measured for hardness in two different places; several castings were made under the actual conditions to let the process stabilize for the designated operating conditions. The results may be condensed into the results for a trial and a position, and for a trial only as shown in Table 8-8.

8-7-2   ANOVA and interpretation of raw data

The primary portion of Table 8-8 contains the variation due to control factors only, since the position and repetitions have been condensed into one trial value. Because this value varies from trial to trial, it is a function of influence of the control factors only. The secondary portion of

TABLE 8-6   Die Cast Component Factors and Levels

Factors		Level 1	Level 2
A	% copper	Lower spec. limit	Upper spec. limit
B	% magnesium	Lower spec. limit	Upper spec. limit
C	% zinc	Lower spec. limit	Upper spec. limit
D	Water cooling	On	Off
E	Air cooling	On	Off
P	Position (on casting)	I	II

TABLE 8-7 Die Cast Component Experiment Tertiary Table

	A	B	A×B	C	D	E	e						
	Column no.							Data (R_B) at position (P)					
Trial no.	1	2	3	4	5	6	7	I			II		
1	1	1	1	1	1	1	1	71	71	72	75	74	75
2	1	1	1	2	2	2	2	72	72	72	71	69	71
3	1	2	2	1	1	2	2	55	55	55	68	67	68
4	1	2	2	2	2	1	1	76	74	74	72	74	74
5	2	1	2	1	2	1	2	78	78	77	78	78	76
6	2	1	2	2	1	2	1	63	69	65	69	73	66
7	2	2	1	1	2	2	1	74	70	72	72	70	74
8	2	2	1	2	1	1	2	75	72	71	73	74	73

Table 8-8 contains the variation due to control and noise factors, since the repetitions have been condensed into one value. Table 8-7, the tertiary table, contains the total variation due to control factors, noise factors, and error. For this reason, three SS_T values may be calculated from the data.

Primary

$$SS_{T_1} = \sum_{i=1}^{c_p}\left(\frac{y_{1_i}^2}{n_{1_i}}\right) - \frac{T^2}{N}$$

where c = number of treatment conditions in table
       c_p = number of data points in primary table

$$SS_{T_1} = \frac{438^2}{6} + \frac{427^2}{6} + \cdots + \frac{438^2}{6} - \frac{3417^2}{48} = 994.15$$

TABLE 8-8 Condensed Experimental Results

Secondary table			Primary table	
	Position totals			Trial
Trial no.	I	II	Trial no.	total
1	214	224	1	438
2	216	211	2	427
3	165	203	3	368
4	224	220	4	444
5	233	232	5	465
6	197	208	6	405
7	216	216	7	432
8	218	220	8	438
Totals	1683	1734	Total	3417

Secondary

$$SS_{T_2} = \sum_{i=1}^{c_s} \left(\frac{y_{2_i}^2}{n_{2_i}} \right) - \frac{T^2}{N}$$

where c_s = number of data points in secondary table

$$SS_{T_2} = \frac{214^2}{3} + \frac{216^2}{3} + \cdots + \frac{220^2}{3} - \frac{3417^2}{48} = 1279.31$$

Tertiary

$$SS_{T_3} = \sum_{i=1}^{N} (y_{3_i}^2) - \frac{T^2}{N}$$

$$SS_{T_3} = 71^2 + 71^2 + \cdots + 73^2 - \frac{3417^2}{48} = 1361.31$$

The variation due to control factors (inner array columns) may be calculated by using the primary table; the total of the column effects must equal SS_{T_1}. The variation due to control factors, noise factors, and control and noise factor interactions may be calculated from the secondary table; the total of all these effects must equal SS_{T_2}. The variation due to control factors, noise factors, control and noise factor interactions, and error may be calculated by using the tertiary table; the total of all these effects must equal SS_{T_3}. The inner OA column effects may be calculated in the manner used in Chap. 3. The noise factor and control and noise factor interactions may be calculated in the ANOVA manner used in Chap. 2. The layout for the % copper and position interaction is shown in Table 8-9. The total in each treatment condition is obtained by summing the data under the specified condition. In this case, trials 1 to 4 (level 1) and position I add up to 819; trials 5 to 8 (level 2) and position I add up to 864; and so on. Each inner OA column by outer array factor interaction may be determined in the same manner.

TABLE 8-9   % Copper and Position Interaction

		Position	
		I	II
% copper	1	819	858
	2	864	876

TABLE 8-10   Die Cast Experiment ANOVA (Raw Data)

Source	SS	ν	V
A	82.69	1	82.69
B	58.52	1	58.52
C	2.52	1	2.52
D	295.02	1	295.02
E	487.69	1	487.69
$A \times B$	58.52	1	58.52
e_1	9.19	1	9.19
SS_{T_1}	994.15	7	
P	54.19	1	54.19
$A \times P$	15.18	1	15.18
$B \times P$	9.19	1	9.19
$C \times P$	38.52	1	38.52
$D \times P$	105.02	1	105.02
$E \times P$	28.52	1	28.52
$A \times B \times P$	28.52	1	28.52
e_2	6.02	1	6.02
SS_{T_2}	1279.31	15	
e_3	82.00	32	2.56
SS_{T_3}	1361.31	47	

$$SS_{A \times \text{Position}} = \sum_{i=1}^{c} \left(\frac{y_i^2}{n_i} \right) - SS_A - SS_{\text{Position}} - \frac{T^2}{N}$$

$$= \frac{819^2 + 864^2 + 858^2 + 876^6}{12} - 82.69 - 54.19 - \frac{3417^2}{48}$$

$$= 15.18$$

The complete ANOVA table for the raw data analysis would appear as in Table 8-10, and one pooled version of the ANOVA table is shown in Table 8-11.

The water cooling by position interaction is statistically significant in this case but since position is a noise factor the only option is to choose the proper level for water cooling. A plot of the average results of the interaction treatment conditions would make this selection clearer. The four combinations of factors D and P are:

D	P	Average RHN$_B$
1	1	66.17
1	2	71.25
2	1	74.08
2	2	73.25

TABLE 8-11 Die Cast Experiment Pooled ANOVA (Raw Data)

Source	SS	ν	V	F	P
A	82.69	1	82.69	9.10#	5.41
D	295.02	1	295.02	32.46#	21.00
E	487.69	1	487.69	53.65#	35.16
$D \times P$	105.02	1	105.02	11.55#	7.05
e_p	390.89	43	9.09		31.38
SS_{T_3}	1361.31	47			100.00

†At least 90% confidence.
‡At least 95% confidence.
#At least 99% confidence.

A plot of the results is shown in Fig. 8-6. Notice, the average hardness of D_2 is higher and more consistent in spite of the noise factor. Using level 2 of factor D is parameter design with respect to component hardness.

The averages of the levels of factors A, D, and E are

$$\overline{A}_1 = 69.88 \qquad \overline{D}_1 = 68.71 \qquad \overline{E}_1 = 74.38$$

$$\overline{A}_2 = 72.50 \qquad \overline{D}_2 = 73.67 \qquad \overline{E}_2 = 68.00$$

The estimated average results when the three control factors are at their better level is

$$\mu_{A_2 D_2 E_1} = \overline{A}_2 + \overline{D}_2 + \overline{E}_1 - 2\overline{T}$$

$$= 72.50 + 73.67 + 74.38 - 2\,(71.19) = 78.18$$

The 90% confidence interval for the population and the confirmation experiment of 15 parts is

$$CI_{pop} = \sqrt{\frac{F_{\alpha;1;\nu_e}V_e}{n_e}} = \sqrt{\frac{2.83(9.09)}{12}} = 1.46$$

$$F_{.10;1;43} = 2.83$$

$$n_e = \frac{N}{1 + \nu_A + \nu_D + \nu_E} = \frac{48}{4} = 12$$

$$\mu_{pop} = 78.18 \pm 1.46$$

$$76.72 < \mu_{pop} < 79.64$$

$$CI_{CE} = \sqrt{F_{\alpha;1;\nu_e}V_e(1/n_e + 1/r)}$$

$$CI_{CE} = \sqrt{2.83(9.09)(1/12 + 1/30)} = 1.73$$

$$r = 30 \text{ (each of the 15 parts will be measured in two places)}$$

$$76.45 < \mu_{CE} < 79.91$$

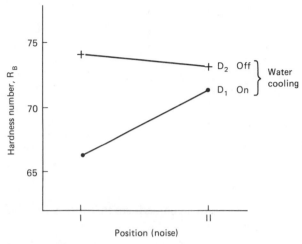

Figure 8-6 Die casting noise interaction.

There are six data points (three parts) in this experiment which were manufactured under the conditions of $A_2D_2E_1$ and had hardness values of 78, 78, 77, 78, 78, and 76. This is an average of 77.50, which is well within the interval for a 30-piece confirmation experiment. This indicates that the confirmation experiment should be successful. If production parts have this average value and variation, the experiment will have provided a substantial improvement.

8-7-3   ANOVA and interpretation of *S/N* Data

Analysis of the *S/N* ratio follows after the raw data from each trial has been transformed. In this case, the data is transformed to a HB *S/N* value which is contained in Table 8-12. An ANOVA is conducted on the *S/N* values as if they were one data point for each trial. The ANOVA

TABLE 8-12   Hardness of *S/N* Transformation

Trial no.	*S/N*, db
1	37.26
2	37.04
3	35.61
4	37.38
5	37.78
6	36.56
7	37.14
8	37.26

TABLE 8-13 S/N ANOVA Summary

Source	SS	v	V
A	0.26	1	0.26
B	0.19	1	0.19
C	0.02	1	0.02
D	0.88	1	0.88
E	1.39	1	1.39
$A \times B$	0.23	1	0.23
e	0.05	1	0.05
T	3.02	7	

table is summarized in Table 8-13 and in a pooled form in Table 8-14. The averages of the factors D and E are

$$\overline{D}_1 = 36.67 \qquad \overline{E}_1 = 37.42$$

$$\overline{D}_2 = 37.34 \qquad \overline{E}_2 = 36.59$$

The estimated mean is

$$\mu_{D_2 E_1} = \overline{D}_2 + \overline{E}_1 - \overline{T} = 37.34 + 37.42 - 37.01 = 37.75$$

The 90% confidence intervals are

$$\text{CI}_{\text{pop}} = \sqrt{\frac{4.06(.15)}{2.67}} \qquad\qquad F_{.10;1;5} = 4.06$$

$$= .48 \qquad\qquad n_e = \frac{8}{3} = 2.67$$

$$\text{CI}_{\text{CE}} = \sqrt{4.06(.15)[1/2.67 + 1/5]} \qquad\qquad r = 5$$

$$= .59$$

Note that the confirmation experiment for S/N was set up as five repetitions each using six hardness values which are collected from a total of

TABLE 8-14 S/N Pooled ANOVA Summary

Source	SS	v	V	F	P
D	0.88	1	0.88	5.87[†]	24.17
E	1.39	1	1.39	9.27[‡]	41.06
e_p	0.75	5	0.15		34.77
T	3.02	7			100.00

[†]At least 90% confidence.
[‡]At least 95% confidence.
[#]At least 99% confidence.

15 castings. One S/N value must come from the transformation of the same number of repetitions as the original experiment for the confidence interval to apply.

The factors in the die casting experiment can then be classified.

Factor	Effects average	Effects variability
A	(2)#	
D	(2)#	(2)†
E	(1)#	(1)‡

NOTE: Numbers in parentheses indicate best level.
†At least 90% confidence.
‡At least 95% confidence.
#At least 99% confidence.

In this case, the same levels of two of the significant factors provide a higher average and reduced variability so nothing has to be compromised. In some situations, the levels of factors which improve the average and improve uniformity may conflict, so a compromise may have to be reached. Also, a compromise may have to occur when multiple responses are considered and the same factor level may cause one response to improve and another to deteriorate.

8-8 Alternative Inner/Outer OA Experiment

This approach to parameter design is the most sophisticated of the three strategies discussed so far. The first approach used repetitions only as a noise source, as in Table 8-1. A second and somewhat more comprehensive approach utilized an outer array to intentionally include noise in the experiment, as in Table 8-2. The third approach uses the same three-level array as the inner and outer array to investigate the presence of nonlinear performance characteristics. The first approach uses no intentional noise factors, the second approach intentionally uses outer noise factors, and the third approach intentionally uses product noise factors.

In new product or process designs the tendency for nonlinear performance may or may not be known, but a strategy to detect this phenomenon and allow parameter design to be done is desired. The nonlinear performance provides the opportunity to make the performance robust against product noises. Assume a product has nonlinear performance with respect to design parameter C as indicated in Fig. 8-7. Three levels of parameter C could detect the nonlinear tendency. However, if the levels of parameter C are held very precisely for the experiment, the S/N ratios may not differ very much. If product noise is forced into the experiment by using plus and minus values of parameter C around the

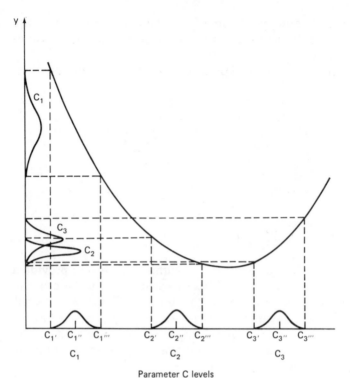

Figure 8-7 Three-level parameter design.

nominal values parameter C, then the S/N ratio would indicate a different variance. So each of the three nominal levels of parameter C (C_1, C_2, C_3) could have three levels within them ($C_{1'}$, $C_{1''}$, $C_{1'''}$, for instance). This experiment emulates the real world situation of having chosen a nominal value for parameter C, but due to manufacturing variation that value is not obtained exactly in every product that is produced. Sometimes the actual value of parameter C is a little lower than nominal and sometimes a little higher than nominal. Determining which nominal is best to reduce sensitivity to product variation is the key. In Fig. 8-7 the best of the three levels of parameter C is the nominal of C_2. Because of the nonlinear performance, the optimum level is between C_2, and C_3. This level would accommodate the largest variation in parameter C without transmitting this variation through to the performance characteristic y.

To structure an experiment to take advantage of this strategy is an extension of the inner/outer OA approach. The three-level inner array is assigned only control factors as before with all of the primes indicating the nominal values for the parameters to be investigated. The outer three-level array is used to designate the variation around the nominals of the control factors; level 1 might be 5% below the nominal; level 2 would be the actual nominal; and level 3 might be 5% above the nominal.

8-8-1   Case study: automobile hood hinge

The automobile hood hinge mechanism mentioned in Sec. 1-4 could have a parameter design study done to develop as consistently functioning a hood as possible. The force to close the hood from the open position, which is automatically held by the mechanism, is an NB characteristic as previously mentioned. The mechanism shown in basic form in Fig. 8-8 is a four-bar linkage with a coil spring located at the pivot point of links A and B. Links A and B have a mechanical stop to establish the angle between those links when the hood is in full open position. The design parameters that are available to investigate are the lengths of the four links ($A, B, C,$ and D), the angle between links A and B, and the coil spring force. Link D could be set at the maximum length that would allow the hood to reach the closed position, since the experiment would include all combinations of the lengths of links A, B, and C. With five factors at three levels, an L18 OA is required for the experimental layout, so the resultant inner/outer array arrangement is as shown in Table 8-15.

The levels for the control factors in the inner array are shown in Table 8-16 and the levels for noise in the outer array are shown in Table 8-17. Obviously, this type of parameter design experiment requires many data points. In many situations the cost and time to prepare hardware preclude this approach. For these reasons, this parameter design ap-

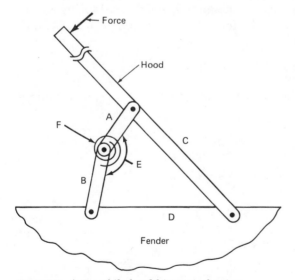

Figure 8-8 Automobile hood hinge mechanism.

TABLE 8-15   Three-Level OA for Parameter Design

proach is used mainly when a mathematical simulation of the product is available. For instance, an equation might be written that would calculate the force required to close the hood based on the lengths of links A, B, C, and D; the angle E between link A and B; and the coil spring load F. The equation would read that the force y is a function of A, B, C, D, E, and F:

$$y = f(A, B, C, D, E, F)$$

A value could then be calculated for each of the 324 possible combina-

TABLE 8-16   Parameter Design Nominal Levels

Control	Nominal levels		
factors	1	2	3
A	7	9 in	11
B	7	9	11
C	9	11	13
D	Dependent on A, B, and C		
E	150°	160°	170°
F	10	20 ft·lb	30

TABLE 8-17   Parameter Design Noise Levels

Control factors	Noise levels		
	1	2	3
A	−1%	Nominal	+1%
B	−1%	Nominal	+1%
C	−1%	Nominal	+1%
D	−1%	Nominal	+1%
E	−5%	Nominal	+5%
F	−10%	Nominal	+10%

tions of the control and noise levels. The value for the $y_{1,1}$ test (first trial in control array and first trial in noise array) would be:

$$y_{1,1} = f(A = 6.93, B = 6.93, C = 8.91, D = 8.91, E = 142.5, F = 9.0)$$

The formula would allow a force value to be predicted for each of the possible combinations of control and noise levels. Then a S/N ratio could be calculated to provide a measure of the variation caused by the noise levels. Two ANOVAs would then be completed to identify which control factors and levels affected the average force and the variation of the force. With this identification complete, the sensitivity of the hood closing force to variation in the hinge mechanism could be reduced. This again would make the hood closing force less sensitive to other sources of noise such as amount of lubricant (outer noise) and wear in the mechanism (inner noise).

8-9   Measurement System Parameter Design

Parameter design can be accomplished on measurement systems by using the signal-to-noise concept. The basic operation of a measurement system is shown in Fig. 8-9. The input or signal into a measurement system is some unknown, but true, value. The diameter of a piston bore is not actually known but provides the signal to a dial bore gauge. Also entering into the measurement system is noise, which may take many forms such as variation in operators, temperature, and vibration. Noise is really any disturbance to the true purpose of the measurement system which is to translate the signal into some scalar value in a precise manner. The accuracy of a measurement system may require compensation or calibration, but the precision (repeatability) is the primary property sought in the device. The output of a measuring system is the observed value. The dial bore gauge mentioned earlier converts the abstract value of diameter of the bore into a concrete length value of inches or millimeters. The true dimension of the bore is unknown, but

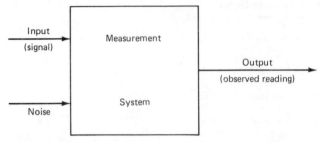

Figure 8-9   Measurement system.

the observed value indicated by the gauge will have to be assumed to be correct because the noise effect is also unknown. There are ways to experiment with measurement systems to assess the repeatability and the impact of noise which require repeated measurement of the same parts in the same position. The average of the repeated measurements tends to cancel out some of the noise effects and provide a better estimate of the true value of the item being measured.

The signal-to-noise concept for quantifying the ability of a measurement system to translate a signal into an observed value is quite useful. The *S/N* value ratios the power of the signal to the power of the noise as they appear in the output.

$$S/N = \frac{\text{power of the signal}}{\text{power of the noise}}$$

In a measurement system parameter design experiment the strategy is modified somewhat to include the signal factor with noise in the outer array. Again, only control factors are assigned to the inner array.

8-9-1   Measurement system example 1

In an automotive thrust washer test, one of the responses of interest is wear of the material which might allow components to shift in location due to a change in thickness of the washer. A gauge, similar in concept to a micrometer or height gauge, is used to detect a change in thickness after a particular thrust washer test. The gauge should be able to detect very small changes in thickness. Since the parts will have to measured before the test and after the test, the precision of the measuring device will have to be very good. The signal factor will be some thickness, the levels of which should take on three equally spaced, known values to span the range of thickness values to be measured by the device. The known values of thickness should be a measurement standard such as jo-blocks or some other trusted standard. The noise factor in this experiment might be different operators used to collect test data. Control

TABLE 8-18 Measurement System Parameter Design

			Inner array					Outer array				
								Signal level				
	A	B	C	D	E	F	G	1		2		3	
			Column no.					(.050)		(.150)		(.250)	
Trial no.	1	2	3	4	5	6	7	N_1	N_2	N_1	N_2	N_1	N_2
1	1	1	1	1	1	1	1	45	53	155	145	248	253
2	1	1	1	2	2	2	2	35	43	145	155	263	258
3	1	2	2	1	1	2	2	48	52	149	151	252	248
4	1	2	2	2	2	1	1						
5	2	1	2	1	2	1	2						
6	2	1	2	2	1	2	1						
7	2	2	1	1	2	2	1						
8	2	2	1	2	1	1	2						

NOTE: First three trials for example only.

factors for the gauge would be such things as gauge-to-washer contact area, gauge-to-washer load, etc. A summary of the factors in the experiment is

Control factors	$A-G$
Noise factor	N (operators 1 and 2)
Signal factor	S (thickness standard)

The experimental arrays would be arranged like those in Table 8-18. Signal factor levels could be such values as .050 in (1.27 mm), .150 in (3.81 mm), and .250 in (6.35 mm), which span the typical values for automotive thrust washers. The recorded data values are the actual readings the measuring device provided when measuring the thickness of the signal level. These values are in thousandths of an inch.

To calculate the S/N value for each trial, the data is rearranged into a noise versus signal matrix. The first trial is represented in Table 8-19. With this arrangement the total variation of these values can be decomposed into variance due to the signal (both linear and quadratic), the noise, the signal and noise interaction, and error. To be useful, however, a gauge should be as linear as possible, so a large amount of variation

TABLE 8-19 Trial 1 Data Summary

	Signal level			
Operator	1	2	3	Totals
1	45	155	248	448
2	53	145	253	451
Totals	98	300	501	899

due to the linear component of the signal is desirable and all other variation should be considered as error. Therefore,

$$SS_T = SS_{S_1} + SS_{S_q} + SS_N + SS_{S \times N} + SS_e$$

$$= \sum_{i=1}^{r} y_i^2 - \frac{T^2}{r} = 45^2 + 53^2 + \cdots + 253^2 + \frac{899^2}{6}$$

$$= 40697.0$$

where r = number of repetitions in a trial.

The linear sum of squares (refer to the formula from Table D.3 of Appendix D and Sec. 4-4) for a three-level factor is

$$SS_{S_1} = \frac{(W_1 S_1 + W_2 S_2 + W_3 S_3)^2}{W_T R}$$

$$= \frac{[-1(98) + 0(300) + 1(501)]^2}{2(2)} = 40{,}602.0$$

where R = number of repetitions within the factor S levels.

The ANOVA summary is shown in Table 8-20. The S/N calculation is based on the nominal-is-best 2 form

$$S/N_{NB_2} = 10 \log \left[\frac{V_m - V_e}{r V_e} \right]$$

In a gauge study, the variance due to S_1 is substituted for the variance due to the mean, and R for r, and the equation then becomes

$$S/N_{measurement} = 10 \log \left[\frac{V_{S_1} - V_e}{R V_e} \right]$$

TABLE 8-20   ANOVA Summary for Trial 1

	Source	SS	ν	V
	S_1	40,602.	1	40,602.00
	S_q			
Error	$\Big\{ \; N$	95.	4	23.75
	$S \times N$			
	$e \; \Big\}$			
	T	40,697.	$\overline{5}$	

For trial 1, $S/N_1 = 10 \log \left[\dfrac{40{,}602 - 23.75}{2(23.75)} \right] = 29.32$

For trial 2, $SS_T = 49{,}157$ $SS_{S_1} = 49{,}062$ $SS_e = 95$

$$S/N_2 = 10 \log \left[\dfrac{49{,}062 - (95/4)}{2(95/4)} \right] = 30.14$$

For trial 3, $SS_T = 40{,}018$ $SS_{S_1} = 40{,}000$ $SS_e = 18$

$$S/N_3 = 10 \log \left[\dfrac{40{,}000 - (18/4)}{2(18/4)} \right] = 36.48$$

Recall that reduced variation will result in a higher S/N value; here reduced variation is equivalent to improved repeatability. The remaining data collected in the trials would be analyzed in the same fashion to find the S/N ratio for that trial. An ANOVA would be done on the eight S/N values to determine which control factors affected the average S/N value and which levels of those factors were more consistent. The intent is to see the measurement system respond in a consistent, linear fashion since the signal, or part to be measured, was structured linearly (three equally spaced values).

The first three S/N values demonstrate some interesting points. The first and second trials have exactly the same repeatability problem from operator to operator when measuring the three test pieces. However, in trial 2 the average reading of the .050 test piece (S_1) is low, the average reading of the .150 test piece (S_2) is on target, and the average reading of the .250 test piece (S_3) is high. Intuitively, it would seem as if trial 2 were a poorer gauge; however, it is really more sensitive to the differences in the signal parts and just needs to be rescaled to provide the proper reading. The respective values indicate trial 2 to be a better measuring system; a higher S/N value is obtained in trial 2. Trial 3 is better yet than trials 1 or 2. Even though not as sensitive as trial 2 (lower slope than trial 2), the repeatability around a point is better.

Small changes in S/N ratio are important. For each 3 db S/N value increases, the error variation is cut in half relative to the variation due to the linear portion of signal (or mean when using the original S/N formula).

8-9-2 Measurement system example 2

In an automatic transmission with electronic shift control logic, a device is needed to sense the output shaft speed, which is proportional to vehicle velocity. When a shift should occur is partly a function of vehicle speed. To sense speed, a transducer converts a varying magnetic field to a voltage with a frequency proportional to vehicle speed. The transducer is a measurement device to detect output shaft speed.

The transducer has a magnetic core around which a coil of conductive wire is wound. The tip of the magnetic core is held close to a rotating wheel which is attached to the output shaft. The wheel is made of magnetically conductive (ferrous) material with notches or teeth on the perimeter as shown in Fig. 8-10. As the notches move past the tip of the magnetic core, the magnetic field within the coil changes, which creates a voltage spike across the ends of the coil wire. The changing speed of the output shaft causes a change in the frequency of the voltage; consequently, the electronic "brain" can use this frequency signal as a measure of vehicle speed. The electronic circuit would like to have frequency only be a function of output shaft speed and have the voltage spike to be consistent over the range of speeds measured.

A designed experiment could be utilized to study such control factors as wire diameter, wire material, number of coil turns, coil diameter, coil length, magnetic core strength, wheel material, tooth shape, and tooth depth. Noise factors could be such things as core to wheel clearance (which may be difficult to hold closely in a transmission assembly), temperature, etc. The signal factor would be output shaft speed at three levels such as 500, 3500, and 6500 rpm. The peak voltage could be one response measured in the experiment and then the same type of *S/N*

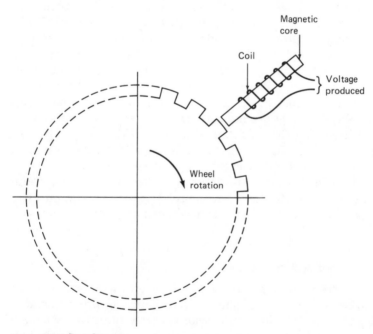

Figure 8-10 Speed sensor.

analysis and ANOVA carried out. This would identify which control factors provided consistent peak voltage at varying speeds, temperature, and clearances.

8-10 Tolerance Design

Tolerance design is utilized when the efforts of parameter design have not proved adequate in reducing variation. In parameter design low cost, widely varying components or factors may be used. If the quality of these components must still be improved to reduce variation to the desired amount, then tolerance design comes into play. In tolerance design the loss function is used to substantiate the increased costs of higher-quality components by lower loss to society.

As an example, the transmission shift point discussion in Sec. 1-9 could be used to demonstrate the principle of tolerance design. The k value for the shift point adjustment was $.0625. The loss function for a nominal-is-best situation is

$$L(y) = k[S_y^2 + (\bar{y} - m)^2]$$

In this case, it is assumed that the distribution of shift speeds will be centered because the variation around the average is the primary concern. Then the loss function simplifies to

$$L(y) = k \, S_y^2$$

If the current variance of shift speeds is 400 (S_y = 20 rpm), then the average loss per transmission is

$$L(y) = \$.0625(400) = \$25.00$$

One of the components which contributes to this variation and subsequent loss is the shift valve spring. As a result of an experiment, the spring is known to have a substantial effect on shift speed, but parameter design approaches with other factors have not reduced the spring effect adequately. A higher-priced spring with less variance would reduce the shift variance to 200 (S_y = 14) at the additional cost of $1.00 per spring (one per transmission). Would this be a wise investment? The new loss is

$$L(y) = \$.0625(200) = \$12.50$$

Spending $1.00 on the spring reduces the loss per transmission by $12.50 according to the loss function with a net gain of $11.50 per transmission. Obviously, this is a substantial reduction in loss and is worthwhile.

The real trick in tolerance design is establishing the relationship of the variance of the component or factor to the variance of the performance characteristic of interest (i.e., spring variance affects shift variance by some relationship). If variance of the component or factor is reduced (quality improved) at some cost, then some reduction in variance of the performance characteristic will be obtained, resulting in a reduced loss per the loss function.

The relationship between factor variances, factors A through F, and the total variance is described by the equation

$$S_T^2 = S_A^2 + S_B^2 + \cdots + S_F^2 + S_e^2$$

The variance indicated here is the variance of the performance characteristic caused by the variance of the factor and not the actual variance of the factor itself. Recall how the variance of a factor is transmitted to the variance of the performance characteristic. The larger variances in this equation are caused by the more influential factors, and the lesser values of variance by the less influential factors.

The loss function uses the value of total variance in calculating an associated loss. If tolerance design is done on a less influential factor, then the total variance will not change that much and the loss function may not substantiate the tolerance design approach for that factor. The factors that contribute more toward the total variance are the ones that will be more effective in utilizing tolerance design.

8-11 Quality Countermeasures

The earliest stages of design and development are the areas of greatest cost reduction in products and processes. Refer again to the three stages of design: system design (SD), parameter design (PD), and tolerance design (TD). Quality may be designed into a product or process by making it robust against all noises (outer, inner, and product), but only at certain stages of the product life cycle.

At the earliest stage of the life cycle, in research and development, SD may be used to improve quality with respect to all noises, as indicated by Table 8-21. One system relative to another may be more robust against noises. Once the system is chosen, then PD may be applied to combat noise effects also. TD can effectively be applied to inner and product noise, but PD should be used for outer noises.

When production engineering starts, the basic system has been selected and so have the nominal values for the design parameters. At this point PD and TD are not very effective against outer and inner noises. Efforts from here through the actual production environment are effective against product noise.

Once the product is sold, no countermeasures are effective. Again, the loss function states that once a product has been shipped, there is some

TABLE 8-21 Quality Countermeasures Against Noises

Area of quality control	Department	Countermeasure	Noises		
			Outer	Inner	Product
Off-line quality control	Research and development	System design	○	○	○
		Parameter design	○	○	○
		Tolerance design	●	○	○
	Production engineering	System design	×	×	○
		Parameter design	×	×	○
		Tolerance design	×	×	○
On-line quality control	Manufacturing	Diagnosis and adjustment of processes	×	×	○
		Forecasting and correction	×	×	○
		Measurement and disposition	×	×	○
	Sales	Service	×	×	×

NOTE: ○ = effective; ● = effective but not recommended; × = impossible.
SOURCE: Genichi Taguchi and Yu-in Wu, *Off-line Quality Control*. Central Japan Quality Control Assoc., Nagaya, 1979, p. 80.

societal loss associated with that product. The bottom line is to start as early as possible in the product life cycle to make as high-quality, low-loss design as possible.

8-12 Steps in Experimentation

This section summarizes the essential steps in beginning an experimental design and following through to interpretation.

Step 1. State the problem to be solved. A clear understanding of the problem by those involved with the experiment is necessary to be able to structure the experiment. The problem statement should be specific, and if multiple responses are involved, that should be noted.

Step 2. Determine the objective of the experiment. This includes the identification of the performance characteristics (preferably measurable) and the level of performance required when the experiment is complete.

Step 3. Determine the measurement method(s). An understanding of how the performance characteristic(s) will be assessed after the experiment is conducted. The measurement system may require a separate experiment to improve the measurement system accuracy and precision.

Step 4. Identify factors which are believed to influence the performance characteristic(s). A group of people associated with the product or process should be utilized. Brainstorming, flowcharting, and/or fishbone charting should be used to aid in the creation of the factors to be investigated. If this is an initial screening experiment, many factors should be included if thought to be influential to the response(s).

Step 5. Separate the factors into control and noise factors. This is a basic parameter design strategy and should be used in place of a cause detection or tolerance design approach.

Step 6. Determine the number of levels and values for all factors. The total degrees of freedom required is a direct function of the number of levels for the factors. For initial screening experiments, the number of levels should be kept low; two levels if possible.

Step 7. Identify control factors that may interact. These interactions use up degrees of freedom also and may dictate the size of the experiment. A strategy may be to select the size of the experiment based only on factors, and if there are any columns available afterward then assign these to an interaction of interest.

Step 8. Draw the required linear graph for the control factors and interactions. The desired factors and interactions may influence the selected orthogonal array.

Step 9. Select orthogonal arrays. The OAs, inner and outer, are a function of the total degrees of freedom required from the factor list and the required linear graph.

Step 10. Assign factors and interactions to columns. The linear graph for an OA may have to be modified to fit the required form. The number of levels in a column may have to be modified at this time also. Both inner and outer arrays will have the same assignment considerations; however, the outer array should not be as complex as the inner array because the outer is noise only which is controlled only in the experiment.

Step 11. Conduct the experiment. Trial data sheets should be constructed to minimize the opportunity for errors in setting the proper levels for the trials (the experiment will no longer be orthogonal and balanced if this mistake should occur). Randomization strategies should be considered during the experiment.

Step 12. Analyze the data. There were many methods discussed for analyzing data: observation method, ranking method, column effects method, ANOVA, S/N ANOVA, average graphs, interaction graphs, etc. If the experiment was unbalanced because of a mistake, the analysis of the data must take this into account or the trial rerun to correct the mistake.

Step 13. Interpret the results. Determine which factors are influential and which are not influential to the performance characteristic(s) of interest.

Step 14.   Select optimum levels of most influential control factors and predict expected results. The influential factors are the only ones necessary to have the level set or controlled. The noninfluential factors should be set at the lowest cost level.

Step 15.   Run a confirmation experiment. This is to demonstrate that the factors and levels chosen for the influential factors do provide the desired results. The noninfluential factors should be set at their economic level during the confirmation run. If results don't turn out as expected, then some important factor(s) may have been left out of the experiment and more screening needs to be done.

Step 16.   Return to step 4 if the objective of the experiment is not met and further optimization is possible with the confirmed factors.

8-13 Summary

This chapter discussed some of the most powerful aspects of the Taguchi methodology, in particular, the parameter design concept. The strategy of making a product or process robust against noises by selecting the proper level for the appropriate parameters is the lowest-cost way of intentionally designing quality into a system. Various methods of designing an experiment with the noise effects taken into account were introduced. One method was to simply repeat the trials and let noise effects show up as they might, another approach utilized a noise array to force noise effects into the experiment, and yet another approach used the variation of the design parameters as a form of noise. All of these approaches will lead to products and processes that are more robust against not only the noise included in the experiment, but against other noises as well. The tolerance design approach of reducing the allowed parameter variation to reduce performance variation should only be utilized after the parameter design approach has proved to be insufficient.

Problem Solutions

Chapter 1

1-1

$$L = k(y - m)^2 = \$2.50$$
$$\$2.50 = (.115 - .125)^2$$
$$k = \$2.50/(.010)^2$$
$$k = 25{,}000$$

1-2

$$L_1 = k[S_1^2 + (\bar{y}_1 - m)^2]$$
$$= 25{,}000\,[.00333^2 + (.125 - .125)^2]$$
$$= \$.28$$
Similarly, $L_2 = \$.07$ and $L_3 = \$.69$

1-3

(a) Reduced variation reduces the associated loss.
(b) Missing the target value is a serious loss compared to hitting the target with increased variation.

Chapter 2

Practice Problem 1

$$SS_T = SS_m + SS_e$$
$$= 15^2 + 16^2 + \cdots + 16^2 = 1822.0$$

$$SS_m = 120^2/8 = 1800.0$$
$$SS_e = 1822.0 - 1800.0 = 22.0$$

Practice Problem 2

$$SS_T = SS_A + SS_e$$
$$= .8^2 + 1.2^2 + \cdots + 2.1^2 + (18.3^2/12) = 2.4825$$
$$SS_A = \frac{4.2^2}{4} + \frac{5.9^2}{4} + \frac{8.2^2}{4} - \frac{18.3^2}{12} = 2.0150$$
$$SS_e = 2.4825 - 2.0150 = 0.4675$$

2-1

$$SS_T = SS_m + SS_e$$
$$= 10^2 + 7^2 + \cdots + 9^2 = 640.0$$
$$SS_m = 66^2/7 = 622.29$$
$$SS_e = 640.0 - 622.29 = 17.71$$

Source	SS	ν	V
m	622.29	1	622.29
e	17.71	6	2.95
T	640.00	7	

2-2

$$SS_T = SS_A + SS_e$$
$$= 12^2 + 15^2 + \cdots + 24^2 - \frac{234^2}{14} = 298.86$$
$$SS_A = \frac{41^2}{3} + \frac{100^2}{5} + \frac{46^2}{4} + \frac{47^2}{2} - \frac{234^2}{14} = 282.69$$
$$SS_e = 298.86 - 282.69 = 16.17$$

Source	SS	ν	V	F
A	282.69	3	94.23	58.17#
e	16.17	10	1.62	
T	298.86	13		

†At least 90% confidence.
‡At least 95% confidence.
#At least 99% confidence.

2-3

$$SS_T = SS_A + SS_B + SS_{A \times B} + SS_e$$
$$= 39.875$$
$$SS_A = (57 - 58)^2/8 = 0.125$$
$$SS_B = (51 - 64)^2/8 = 21.125$$
$$SS_{A \times B} = (63 - 52)^2/8 = 15.125$$
$$SS_e = 39.875 - 0.125 - 21.125 - 15.125 = 3.500$$

Source	SS	ν	V	F
A	0.125	1	0.125	
B	21.125	1	21.125	29.1*
$A \times B$	15.125	1	15.125	20.9*
e	3.500	4	0.875	
T	39.875	7		

†At least 90% confidence.
‡At least 95% confidence.
*At least 99% confidence.

Chapter 3

3-1

Column 2

3-2

Several possibilities exist:

L8 (resolution 1)

Column no.	1	2	3	4	5	6	7
Option 1	C	D	$C \times D$	E	$C \times E$	A	B
Option 2	D	E	A	C	$C \times D$	$C \times E$	B
Option 3	A	B	C	D	E	$C \times E$	$C \times D$
L16 (resolution 3)							
Column no.	1	2	4	8	11	12	15
	A	B	C	D	$C \times E$	$C \times D$	E

3-3

Again several possibilities exist, but two options are listed for an L16 OA assignment.

Column no.	Factors and interactions		Column no.	Factors and interactions	
1	B	C	9	$B \times C$	F
2	D	B	10	F	$B \times D$
3	$B \times D$	$B \times C$	11	$B \times F$	$B \times F$
4	E	A	12	$C \times E$	$A \times D$
5	$B \times E$	$A \times C$	13	$A \times D$	$B \times E$
6	$D \times E$	$A \times B$	14	$A \times B$	$C \times E$
7	$A \times C$	$D \times E$	15	A	E
8	C	D			

Chapter 4

4-1

Again several possibilities exist depending on what the two indicator columns are. The first option reflects using columns 2 and 4 as indicators and the second option uses columns 2 and 6.

Trial no.	Level no. (option 1)	Level no. (option 2)
1	1	1
2	2	2
3	3	4
4	4	3
5	1	1
6	2	2
7	3	4
8	4	3

4-2

$$SS_A = \frac{7^2}{2} + \frac{11^2}{2} + \frac{16^2}{2} + \frac{19^2}{2} - \frac{53^2}{8} = 42.375$$

4-3

$$SS_B = \frac{7^2}{2} + \frac{3^2}{4} + \frac{13^2}{2} - \frac{23^2}{8} = 45.125$$

$$SS_e = (1 - 2)^2/4 = .25$$

4-4

$$SS_{A_1} = [-3(9) - 1(3) + 1(12) + 3(19)]^2/2(20)$$
$$= 38.025$$
$$SS_{A_q} = [+1(9) - 1(3) - 1(12) + 1(19)]^2/2(4)$$
$$= 21.125$$
$$SS_{A_c} = [-1(9) + 3(3) - 3(12) + 1(19)]^2/2(20)$$
$$= 7.225$$

Chapter 7

7-1

Source	SS	v	V	F
A	4.51	1	4.51	21.48#
B	.01	1	.01	
C	.11	1	.11	
D	.11	1	.11	
E	.11	1	.11	
F	.01	1	.01	
G	.11	1	.11	
e	14.92	72	.21	
T	19.89	79		

†At least 90% confidence.
‡At least 95% confidence.
#At least 99% confidence.

7-2

Source	SS	ν	V	F
A	14.36	3	4.79	5.10#
B	10.68	3	3.56	3.79‡
C	1.65	3	.55	
D	.50	3	.17	
E	.92	3	.31	
F	.92	3	.31	
G	.50	3	.17	
e	90.45	96	.94	
T	120.00	117		

†At least 90% confidence.
‡At least 95% confidence.
#At least 99% confidence.

Two-Level Orthogonal Arrays

L4 Array*

Trial no.	Column no.		
	1	2	3
1	1	1	1
2	1	2	2
3	2	1	2
4	2	2	1

**L4 Triangular Table
(Interactions)**

Column no.	Column no.	
	2	3
1	3	2
2		1

*Two-level arrays from Genichi Taguchi and Yu-in Wu, *Off-Line Quality Control*, Central Japan Quality Control Association, Nagaya, 1979, pp. 103–107.

L4 LINEAR GRAPH

1 •————3————• 2

• Main effect (factor)

— Interaction

L8 Array

	Column no.						
Trial no.	1	2	3	4	5	6	7
1	1	1	1	1	1	1	1
2	1	1	1	2	2	2	2
3	1	2	2	1	1	2	2
4	1	2	2	2	2	1	1
5	2	1	2	1	2	1	2
6	2	1	2	2	1	2	1
7	2	2	1	1	2	2	1
8	2	2	1	2	1	1	2

L8 Triangular Table (Interactions)

	Column no.					
Column no.	2	3	4	5	6	7
1	3	2	5	4	7	6
2	—	1	6	7	4	5
3	—	—	7	6	5	4
4	—	—	—	1	2	3
5	—	—	—	—	3	2
6	—	—	—	—	—	1

L8 LINEAR GRAPHS

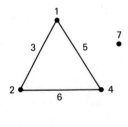

(a)

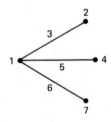

(b)

L12 Array

Trial no.	Column no.										
	1	2	3	4	5	6	7	8	9	10	11
1	1	1	1	1	1	1	1	1	1	1	1
2	1	1	1	1	1	2	2	2	2	2	2
3	1	1	2	2	2	1	1	1	2	2	2
4	1	2	1	2	2	1	2	2	1	1	2
5	1	2	2	1	2	2	1	2	1	2	1
6	1	2	2	2	1	2	2	1	2	1	1
7	2	1	2	2	1	1	2	2	1	2	1
8	2	1	2	1	2	2	2	1	1	1	2
9	2	1	1	2	2	2	1	2	2	1	1
10	2	2	2	1	1	1	1	2	2	1	2
11	2	2	1	2	1	2	1	1	1	2	2
12	2	2	1	1	2	1	2	1	2	2	1

CAUTION. An interaction of two columns is confounded with the remaining columns; assign only main effects to this array.

L16 Array

Trial no.	Column no.														
	1	2	3	4	5	6	7	8	9	10	11	12	13	14	15
1	1	1	1	1	1	1	1	1	1	1	1	1	1	1	1
2	1	1	1	1	1	1	1	2	2	2	2	2	2	2	2
3	1	1	1	2	2	2	2	1	1	1	1	2	2	2	2
4	1	1	1	2	2	2	2	2	2	2	2	1	1	1	1
5	1	2	2	1	1	2	2	1	1	2	2	1	1	2	2
6	1	2	2	1	1	2	2	2	2	1	1	2	2	1	1
7	1	2	2	2	2	1	1	1	1	2	2	2	2	1	1
8	1	2	2	2	2	1	1	2	2	1	1	1	1	2	2
9	2	1	2	1	2	1	2	1	2	1	2	1	2	1	2
10	2	1	2	1	2	1	2	2	1	2	1	2	1	2	1
11	2	1	2	2	1	2	1	1	2	1	2	2	1	2	1
12	2	1	2	2	1	2	1	2	1	2	1	1	2	1	2
13	2	2	1	1	2	2	1	1	2	2	1	1	2	2	1
14	2	2	1	1	2	2	1	2	1	1	2	2	1	1	2
15	2	2	1	2	1	1	2	1	2	2	1	2	1	1	2
16	2	2	1	2	1	1	2	2	1	1	2	1	2	2	1

L16 Triangular Table (Interactions)

Column no.	2	3	4	5	6	7	8	9	10	11	12	13	14	15
1	3	2	5	4	7	6	9	8	11	10	13	12	15	14
2	—	1	6	7	4	5	10	11	8	9	14	15	12	13
3	—	—	7	6	5	4	11	10	9	8	15	14	13	12
4	—	—	—	1	2	3	12	13	14	15	8	9	10	11
5	—	—	—	—	3	2	13	12	15	14	9	8	11	10
6	—	—	—	—	—	1	14	15	12	13	10	11	8	9
7	—	—	—	—	—	—	15	14	13	12	11	10	9	8
8	—	—	—	—	—	—	—	1	2	3	4	5	6	7
9	—	—	—	—	—	—	—	—	3	2	5	4	7	6
10	—	—	—	—	—	—	—	—	—	1	6	7	4	5
11	—	—	—	—	—	—	—	—	—	—	7	6	5	4
12	—	—	—	—	—	—	—	—	—	—	—	1	2	3
13	—	—	—	—	—	—	—	—	—	—	—	—	3	2
14	—	—	—	—	—	—	—	—	—	—	—	—	—	1

L16 Linear Graphs

L16 LINEAR GRAPHS

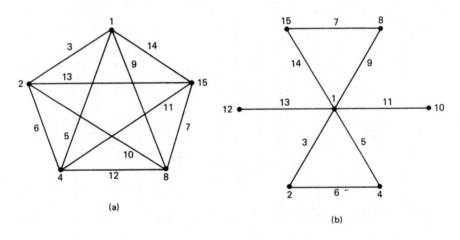

(a)

(b)

L16 Linear Graphs (Continued)

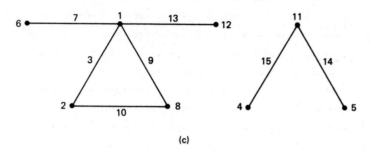

(c)

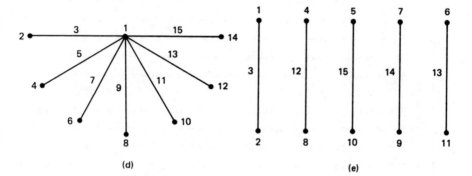

(d)

(e)

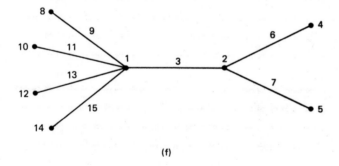

(f)

L32 Array

Column no.

Trial no.	1	2	3	4	5	6	7	8	9	10	11	12	13	14	15	16	17	18	19	20	21	22	23	24	25	26	27	28	29	30	31
1	1	1	1	1	1	1	1	1	1	1	1	1	1	1	1	1	1	1	1	1	1	1	1	1	1	1	1	1	1	1	1
2	1	1	1	1	1	1	1	1	1	1	1	1	1	1	1	2	2	2	2	2	2	2	2	2	2	2	2	2	2	2	2
3	1	1	1	1	1	1	1	2	2	2	2	2	2	2	2	1	1	1	1	1	1	1	1	2	2	2	2	2	2	2	2
4	1	1	1	1	1	1	1	2	2	2	2	2	2	2	2	2	2	2	2	2	2	2	2	1	1	1	1	1	1	1	1
5	1	1	1	2	2	2	2	1	1	1	1	2	2	2	2	1	1	1	1	2	2	2	2	1	1	1	1	2	2	2	2
6	1	1	1	2	2	2	2	1	1	1	1	2	2	2	2	2	2	2	2	1	1	1	1	2	2	2	2	1	1	1	1
7	1	1	1	2	2	2	2	2	2	2	2	1	1	1	1	1	1	1	1	2	2	2	2	2	2	2	2	1	1	1	1
8	1	1	1	2	2	2	2	2	2	2	2	1	1	1	1	2	2	2	2	1	1	1	1	1	1	1	1	2	2	2	2
9	1	2	2	1	1	2	2	1	1	2	2	1	1	2	2	1	1	2	2	1	1	2	2	1	1	2	2	1	1	2	2
10	1	2	2	1	1	2	2	1	1	2	2	1	1	2	2	2	2	1	1	2	2	1	1	2	2	1	1	2	2	1	1
11	1	2	2	1	1	2	2	2	2	1	1	2	2	1	1	1	1	2	2	1	1	2	2	2	2	1	1	2	2	1	1
12	1	2	2	1	1	2	2	2	2	1	1	2	2	1	1	2	2	1	1	2	2	1	1	1	1	2	2	1	1	2	2
13	1	2	2	2	2	1	1	1	1	2	2	2	2	1	1	1	1	2	2	2	2	1	1	1	1	2	2	2	2	1	1
14	1	2	2	2	2	1	1	1	1	2	2	2	2	1	1	2	2	1	1	1	1	2	2	2	2	1	1	1	1	2	2
15	1	2	2	2	2	1	1	2	2	1	1	1	1	2	2	1	1	2	2	2	2	1	1	2	2	1	1	1	1	2	2
16	1	2	2	2	2	1	1	2	2	1	1	1	1	2	2	2	2	1	1	1	1	2	2	1	1	2	2	2	2	1	1
17	2	1	2	1	2	1	2	1	2	1	2	1	2	1	2	1	2	1	2	1	2	1	2	1	2	1	2	1	2	1	2
18	2	1	2	1	2	1	2	1	2	1	2	1	2	1	2	2	1	2	1	2	1	2	1	2	1	2	1	2	1	2	1
19	2	1	2	1	2	1	2	2	1	2	1	2	1	2	1	1	2	1	2	1	2	1	2	2	1	2	1	2	1	2	1
20	2	1	2	1	2	1	2	2	1	2	1	2	1	2	1	2	1	2	1	2	1	2	1	1	2	1	2	1	2	1	2
21	2	1	2	2	1	2	1	1	2	1	2	2	1	2	1	1	2	1	2	2	1	2	1	1	2	1	2	2	1	2	1
22	2	1	2	2	1	2	1	1	2	1	2	2	1	2	1	2	1	2	1	1	2	1	2	2	1	2	1	1	2	1	2
23	2	1	2	2	1	2	1	2	1	2	1	1	2	1	2	1	2	1	2	2	1	2	1	2	1	2	1	1	2	1	2
24	2	1	2	2	1	2	1	2	1	2	1	1	2	1	2	2	1	2	1	1	2	1	2	1	2	1	2	2	1	2	1
25	2	2	1	1	2	2	1	1	2	2	1	1	2	2	1	1	2	2	1	1	2	2	1	1	2	2	1	1	2	2	1
26	2	2	1	1	2	2	1	1	2	2	1	1	2	2	1	2	1	1	2	2	1	1	2	2	1	1	2	2	1	1	2
27	2	2	1	1	2	2	1	2	1	1	2	2	1	1	2	1	2	2	1	1	2	2	1	2	1	1	2	2	1	1	2
28	2	2	1	1	2	2	1	2	1	1	2	2	1	1	2	2	1	1	2	2	1	1	2	1	2	2	1	1	2	2	1
29	2	2	1	2	1	1	2	1	2	2	1	2	1	1	2	1	2	2	1	2	1	1	2	1	2	2	1	2	1	1	2
30	2	2	1	2	1	1	2	1	2	2	1	2	1	1	2	2	1	1	2	1	2	2	1	2	1	1	2	1	2	2	1
31	2	2	1	2	1	1	2	2	1	1	2	1	2	2	1	1	2	2	1	2	1	1	2	2	1	1	2	1	2	2	1
32	2	2	1	2	1	1	2	2	1	1	2	1	2	2	1	2	1	1	2	1	2	2	1	1	2	2	1	2	1	1	2

SOURCE: Reprinted with permission of the American Supplier Institute, Inc.

L32 Triangular Table (Interactions)

Column no.

Column no.	2	3	4	5	6	7	8	9	10	11	12	13	14	15	16	17	18	19	20	21	22	23	24	25	26	27	28	29	30	31
1	3	2	5	4	7	6	9	8	11	10	13	12	15	14	17	16	19	18	21	20	23	22	25	24	27	26	29	28	31	30
2		1	6	7	4	5	10	11	8	9	14	15	12	13	18	19	16	17	22	23	20	21	26	27	24	25	30	31	28	29
3			7	6	5	4	11	10	9	8	15	14	13	12	19	18	17	16	23	22	21	20	27	26	25	24	31	30	29	28
4				1	2	3	12	13	14	15	8	9	10	11	20	21	22	23	16	17	18	19	28	29	30	31	24	25	26	27
5					3	2	13	12	15	14	9	8	11	10	21	20	23	22	17	16	19	18	29	28	31	30	25	24	27	26
6						1	14	15	12	13	10	11	8	9	22	23	20	21	18	19	16	17	30	31	28	29	26	27	24	25
7							15	14	13	12	11	10	9	8	23	22	21	20	19	18	17	16	31	30	29	28	27	26	25	24
8								1	2	3	4	5	6	7	24	25	26	27	28	29	30	31	16	17	18	19	20	21	22	23
9									3	2	5	4	7	6	25	24	27	26	29	28	31	30	17	16	19	18	21	20	23	22
10										1	6	7	4	5	26	27	24	25	30	31	28	29	18	19	16	17	22	23	20	21
11											7	6	5	4	27	26	25	24	31	30	29	28	19	18	17	16	23	22	21	20
12												1	2	3	28	29	30	31	24	25	26	27	20	21	22	23	16	17	18	19
13													3	2	29	28	31	30	25	24	27	26	21	20	23	22	17	16	19	18
14														1	30	31	28	29	26	27	24	25	22	23	20	21	18	19	16	17
15															31	30	29	28	27	26	25	24	23	22	21	20	19	18	17	16
16																1	2	3	4	5	6	7	8	9	10	11	12	13	14	15
17																	3	2	5	4	7	6	9	8	11	10	13	12	15	14
18																		1	6	7	4	5	10	11	8	9	14	15	12	13
19																			7	6	5	4	11	10	9	8	15	14	13	12
20																				1	2	3	12	13	14	15	8	9	10	11
21																					3	2	13	12	15	14	9	8	11	10
22																						1	14	15	12	13	10	11	8	9
23																							15	14	13	12	11	10	9	8
24																								1	2	3	4	5	6	7
25																									3	2	5	4	7	6
26																										1	6	7	4	5
27																											7	6	5	4
28																												1	2	3
29																													3	2
30																														1

SOURCE: Reprinted with permission of the American Supplier Institute, Inc.

L32 Linear Graphs*

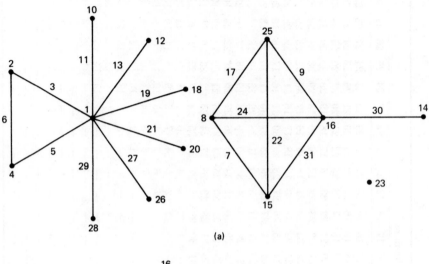

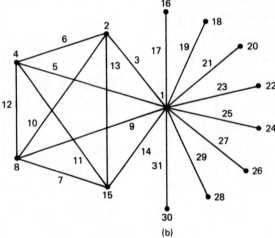

L32 LINEAR GRAPHS

L32 Linear Graphs (Continued)

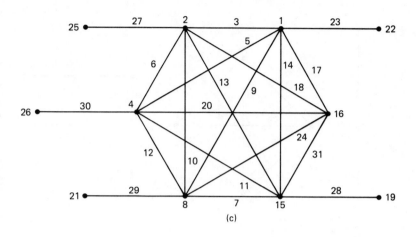

(c)

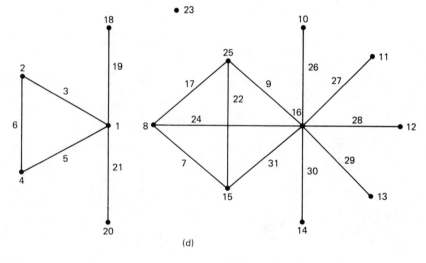

(d)

L32 Linear Graphs (Continued)

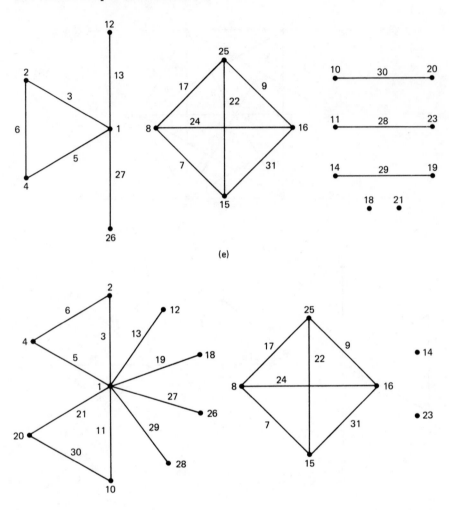

(e)

(f)

L32 Linear Graphs (Continued)

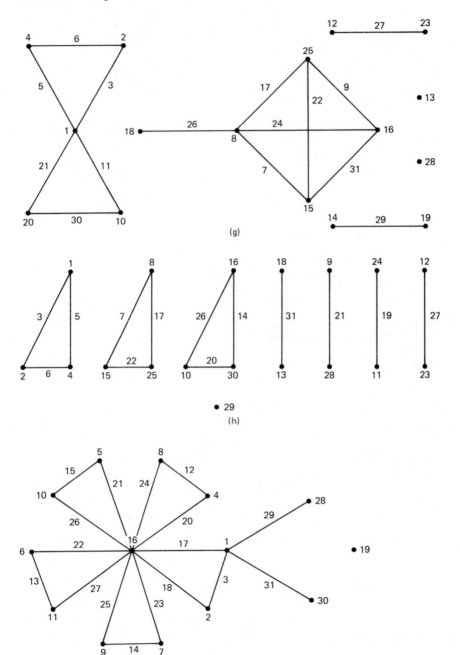

(g)

(h)

• 29

(i)

L32 Linear Graphs (Continued)

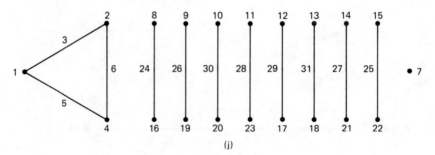

(j)

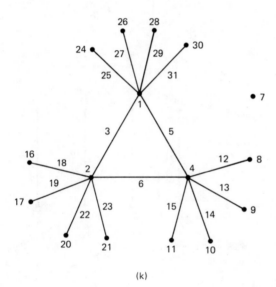

(k)

L32 Linear Graphs (Continued)

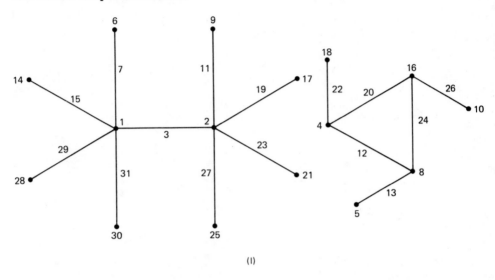

(l)

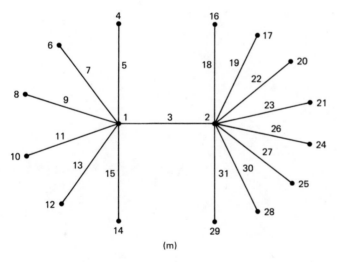

(m)

Three-Level Orthogonal Arrays

L9 Array*

Trial no.	Column no.			
	1	2	3	4
1	1	1	1	1
2	1	2	2	2
3	1	3	3	3
4	2	1	2	3
5	2	2	3	1
6	2	3	1	2
7	3	1	3	2
8	3	2	1	3
9	3	3	2	1

L9 Linear Graph

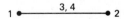

1 •————3, 4————• 2

---

*Three-level arrays from Genichi Taguchi and Yu-in Wu, *Off-Line Quality Control*, Central Japan Quality Control Association, Nagaya, 1979, pp. 108–110.

L18 Array

Trial no.	Column no.							
	1	2	3	4	5	6	7	8
1	1	1	1	1	1	1	1	1
2	1	1	2	2	2	2	2	2
3	1	1	3	3	3	3	3	3
4	1	2	1	1	2	2	3	3
5	1	2	2	2	3	3	1	1
6	1	2	3	3	1	1	2	2
7	1	3	1	2	1	3	2	3
8	1	3	2	3	2	1	3	1
9	1	3	3	1	3	2	1	2
10	2	1	1	3	3	2	2	1
11	2	1	2	1	1	3	3	2
12	2	1	3	2	2	1	1	3
13	2	2	1	2	3	1	3	2
14	2	2	2	3	1	2	1	3
15	2	2	3	1	2	3	2	1
16	2	3	1	3	2	3	1	2
17	2	3	2	1	3	1	2	3
18	2	3	3	2	1	2	3	1

L18 Linear Graph

1 •————————• 2

Two-way ANOVA of columns 1 and 2 for *only* 1 interaction which has 2 d.f. associated with it.

L27 Array

Trial no.					Column no.								
	1	2	3	4	5	6	7	8	9	10	11	12	13
1	1	1	1	1	1	1	1	1	1	1	1	1	1
2	1	1	1	1	2	2	2	2	2	2	2	2	2
3	1	1	1	1	3	3	3	3	3	3	3	3	3
4	1	2	2	2	1	1	1	2	2	2	3	3	3
5	1	2	2	2	2	2	2	3	3	3	1	1	1
6	1	2	2	2	3	3	3	1	1	1	2	2	2
7	1	3	3	3	1	1	1	3	3	3	2	2	2
8	1	3	3	3	2	2	2	1	1	1	3	3	3
9	1	3	3	3	3	3	3	2	2	2	1	1	1
10	2	1	2	3	1	2	3	1	2	3	1	2	3
11	2	1	2	3	2	3	1	2	3	1	2	3	1
12	2	1	2	3	3	1	2	3	1	2	3	1	2
13	2	2	3	1	1	2	3	2	3	1	3	1	2
14	2	2	3	1	2	3	1	3	1	2	1	2	3
15	2	2	3	1	3	1	2	1	2	3	2	3	1
16	2	3	1	2	1	2	3	3	1	2	2	3	1
17	2	3	1	2	2	3	1	1	2	3	3	1	2
18	2	3	1	2	3	1	2	2	3	1	1	2	3
19	3	1	3	2	1	3	2	1	3	2	1	3	2
20	3	1	3	2	2	1	3	2	1	3	2	1	3
21	3	1	3	2	3	2	1	3	2	1	3	2	1
22	3	2	1	3	1	3	2	2	1	3	3	2	1
23	3	2	1	3	2	1	3	3	2	1	1	3	2
24	3	2	1	3	3	2	1	1	3	2	2	1	3
25	3	3	2	1	1	3	2	3	2	1	2	1	3
26	3	3	2	1	2	1	3	1	3	2	3	2	1
27	3	3	2	1	3	2	1	2	1	3	1	3	2

L27 Triangular Table (Interactions)

Column no.						Column no.						
	2	3	4	5	6	7	8	9	10	11	12	13
1	3*	2	2	6	5	5	9	8	8	12	11	11
	4*	4	3	7	7	6	10	10	9	13	13	12
2	—	1	1	8	9	10	5	6	7	5	6	7
	—	4	3	11	12	13	11	12	13	8	9	10
3	—	—	1	9	10	8	7	5	6	6	7	5
	—	—	2	13	11	12	12	13	11	10	8	9
4	—	—	—	10	8	9	6	7	5	7	5	6
	—	—	—	12	13	11	13	11	12	9	10	8
5	—	—	—	—	1	1	2	3	4	2	4	3
	—	—	—	—	7	6	11	13	12	8	10	9
6	—	—	—	—	—	1	4	2	3	3	2	4
	—	—	—	—	—	5	3	12	11	10	9	8
7	—	—	—	—	—	—	3	4	2	4	3	2
	—	—	—	—	—	—	12	11	13	9	8	10
8	—	—	—	—	—	—	—	1	1	2	3	4
	—	—	—	—	—	—	—	10	9	5	7	6
9	—	—	—	—	—	—	—	—	1	4	2	3
	—	—	—	—	—	—	—	—	8	7	6	5
10	—	—	—	—	—	—	—	—	—	3	4	2
	—	—	—	—	—	—	—	—	—	6	7	7
11	—	—	—	—	—	—	—	—	—	—	1	1
	—	—	—	—	—	—	—	—	—	—	13	12
12	—	—	—	—	—	—	—	—	—	—	—	1
	—	—	—	—	—	—	—	—	—	—	—	11

*Pairs of numbers indicate columns that contain the total interaction.

L27 Linear Graphs

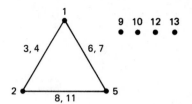

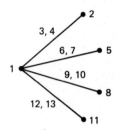

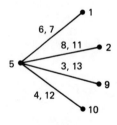

Design and Analysis Tables

TABLE D.1 Resolution of Experimental Arrays

Number of trials	Number of levels	Number of factors	Resolution
4	2	1	4
4	2	2	4
4	2	3	1
8	2	1–3	4
8	2	4	2
8	2	5–7	1
9	3	1	4
9	3	2	4
9	3	3	1
9	3	4	1
16	2	1–4	4
16	2	5	3
16	2	6–8	2
16	2	9–15	1
18	3	1–8	1
27	3	1–3	4
27	3	4–13	1
32	2	1–5	4
32	2	6	3
32	2	7–16	2
32	2	17–31	1

Resolution Code:

1 = Only main factors can be estimated (main effects are all confounded with any order interaction). Low resolution.

2 = All main factors and groups of two-factor interactions (all main effects are unconfounded with two-factor interactions; two-factor interactions are confounded with each other).

3 = All main factors and two-factor interactions can be estimated (all main effects and two-factor interactions are unconfounded).

4 = All main effects and all interactions can be estimated. High resolution.

TABLE D.2 Resolution Two-Column Assignments

Factor	L8	L16	L32
A	1	1	1
B	2	2	2
C	4	4	4
D	7	7	7
E		8	8
F		11	11
G		13	13
H		14	14
I			17
J			18
K			20
L			23
M			24
N			27
O			29
P			30

TABLE D.3 Polynomial Decomposition

Coefficient	$K = 2$	$K = 3$		$K = 4$			$K = 5$			
	L	L	Q	L	Q	C	L	Q	C	F
W_1	−1	−1	1	−3	1	−1	−2	2	−1	1
W_2	1	0	−2	−1	−1	3	−1	−1	2	−4
W_3		1	1	1	−1	−3	0	−2	0	6
W_4				3	1	1	1	−1	−2	−4
W_5							2	2	1	1
W_T	2	2	6	20	4	20	10	14	10	70

$$SS_A = \frac{(W_1 \times A_1 + \cdots + W_K \times A_K)^2}{W_T \times R} \qquad \text{Sum of squares}$$

NOTE: R = number of replications within a level (must be equal); L = linear; Q = quadratic; C = cubic; F = fourth.

TABLE D.4   F Values

		$F_{.10;\nu_1;\nu_2}$		90% confidence[†]							
		Degrees of freedom for the numerator (ν_1)									
		1	2	3	4	5	6	7	8	9	10
Degrees of freedom for the denominator (ν_2)	1	39.9	49.5	53.6	55.8	57.2	58.2	58.9	59.4	59.9	60.2
	2	8.53	9.00	9.16	9.24	9.29	9.33	9.35	9.37	9.38	9.39
	3	5.54	5.46	5.39	5.34	5.31	5.28	5.27	5.25	5.24	5.23
	4	4.54	4.32	4.19	4.11	4.05	4.01	3.98	3.95	3.94	3.92
	5	4.06	3.78	3.62	3.52	3.45	3.40	3.37	3.34	3.32	3.30
	6	3.78	3.46	3.29	3.18	3.11	3.05	3.01	2.98	2.96	2.94
	7	3.59	3.26	3.07	2.96	2.88	2.83	2.78	2.75	2.72	2.70
	8	3.46	3.11	2.92	2.81	2.73	2.67	2.62	2.59	2.56	2.54
	9	3.36	3.01	2.81	2.69	2.61	2.55	2.51	2.47	2.44	2.42
	10	3.28	2.92	2.73	2.61	2.52	2.46	2.41	2.38	2.35	2.32
	11	3.23	2.86	2.66	2.54	2.45	2.39	2.34	2.30	2.27	2.25
	12	3.18	2.81	2.61	2.48	2.39	2.33	2.28	2.24	2.21	2.19
	13	3.14	2.76	2.56	2.43	2.35	2.28	2.23	2.20	2.16	2.14
	14	3.10	2.73	2.52	2.39	2.31	2.24	2.19	2.15	2.12	2.10
	15	3.07	2.70	2.49	2.36	2.27	2.21	2.16	2.12	2.09	2.06
	16	3.05	2.67	2.46	2.33	2.24	2.18	2.13	2.09	2.06	2.03
	17	3.03	2.64	2.44	2.31	2.22	2.15	2.10	2.06	2.03	2.00
	18	3.01	2.62	2.42	2.29	2.20	2.13	2.08	2.04	2.00	1.98
	19	2.99	2.61	2.40	2.27	2.18	2.11	2.06	2.02	1.98	1.96
	20	2.97	2.59	2.38	2.25	2.16	2.09	2.04	2.00	1.96	1.94
	22	2.95	2.56	2.35	2.22	2.13	2.06	2.01	1.97	1.93	1.90
	24	2.93	2.54	2.33	2.19	2.10	2.04	1.98	1.94	1.91	1.88
	26	2.91	2.52	2.31	2.17	2.08	2.01	1.96	1.92	1.88	1.86
	28	2.89	2.50	2.29	2.16	2.06	2.00	1.94	1.90	1.87	1.84
	30	2.88	2.49	2.28	2.14	2.05	1.98	1.93	1.88	1.85	1.82
	40	2.84	2.44	2.23	2.09	2.00	1.93	1.87	1.83	1.79	1.76
	50	2.81	2.41	2.20	2.06	1.97	1.90	1.84	1.80	1.76	1.73
	60	2.79	2.39	2.18	2.04	1.95	1.87	1.82	1.77	1.74	1.71
	80	2.77	2.37	2.15	2.02	1.92	1.85	1.79	1.75	1.71	1.68
	100	2.76	2.36	2.14	2.00	1.91	1.83	1.78	1.73	1.70	1.66
	200	2.73	2.33	2.11	1.97	1.88	1.80	1.75	1.70	1.66	1.63
	500	2.72	2.31	2.10	1.96	1.86	1.79	1.73	1.68	1.64	1.61
	∞	2.71	2.30	2.08	1.94	1.85	1.77	1.72	1.67	1.63	1.60

TABLE D.4 F Values (Continued)

$$F_{.05;\nu_1;\nu_2} \qquad 95\% \text{ confidence}^{\ddagger}$$

ν_2	Degrees of freedom for the numerator (ν_1)									
	1	2	3	4	5	6	7	8	9	10
1	161	200	216	225	230	234	237	239	241	242
2	18.5	19.0	19.2	19.2	19.3	19.3	19.4	19.4	19.4	19.4
3	10.1	9.55	9.28	9.12	9.01	8.94	8.89	8.85	8.81	8.79
4	7.71	6.94	6.59	6.39	6.26	6.16	6.09	6.04	6.00	5.96
5	6.61	5.79	5.41	5.19	5.05	4.95	4.88	4.82	4.77	4.74
6	5.99	5.14	4.76	4.53	4.39	4.28	4.21	4.15	4.10	4.06
7	5.59	4.74	4.35	4.12	3.97	3.87	3.79	3.73	3.68	3.64
8	5.32	4.46	4.07	3.84	3.69	3.58	3.50	3.44	3.39	3.35
9	5.12	4.26	3.86	3.63	3.48	3.37	3.29	3.23	3.18	3.14
10	4.96	4.10	3.71	3.48	3.33	3.22	3.14	3.07	3.02	2.98
11	4.84	2.98	3.50	3.36	3.20	3.01	2.95	2.90	2.85	2.82
12	4.75	3.89	3.49	3.26	3.11	3.00	2.91	2.85	2.80	2.75
13	4.67	3.81	3.41	3.18	3.03	2.92	2.83	2.77	2.71	2.67
14	4.60	3.74	3.34	3.11	2.96	2.85	2.76	2.70	2.65	2.60
15	4.54	3.68	3.29	3.06	2.90	2.79	2.71	2.64	2.59	2.54
16	4.49	3.63	3.24	3.01	2.85	2.74	2.66	2.59	2.54	2.49
17	4.45	3.59	3.20	2.96	2.81	2.70	2.61	2.55	2.49	2.45
18	4.41	3.55	3.16	2.93	2.77	2.66	2.58	2.51	2.46	2.41
19	4.38	3.52	3.13	2.90	2.74	2.63	2.54	2.48	2.42	2.38
20	4.35	3.49	3.10	2.87	2.71	2.60	2.51	2.45	2.39	2.35
21	4.32	3.47	3.07	2.82	2.68	2.57	2.49	2.42	2.37	2.32
22	4.30	3.44	3.05	2.84	2.66	2.55	2.46	2.40	2.34	2.30
23	4.28	3.42	3.03	2.80	2.64	2.53	2.44	2.37	2.32	2.27
24	4.26	3.40	3.01	2.78	2.62	2.51	2.42	2.36	2.30	2.25
25	4.24	3.39	2.99	2.76	2.60	2.49	2.40	2.34	2.28	2.24
26	4.23	3.37	2.98	2.74	2.59	2.47	2.39	2.32	2.27	2.22
27	4.21	3.35	2.96	2.73	2.57	2.46	2.37	2.31	2.25	2.20
28	4.20	3.34	2.95	2.71	2.56	2.45	2.36	2.29	2.24	2.19
29	4.18	3.33	2.93	2.70	2.55	2.43	2.35	2.28	2.22	2.18
30	4.17	3.32	2.92	2.69	2.53	2.42	2.33	2.27	2.21	2.16
32	4.15	3.29	2.90	2.67	2.51	2.40	2.31	2.24	2.19	2.14
34	4.13	3.28	2.88	2.65	2.49	2.38	2.29	2.23	2.17	2.12
36	4.11	3.26	2.87	2.63	2.48	2.36	2.28	2.21	2.15	2.11
38	4.10	3.24	2.85	2.62	2.46	2.35	2.26	2.19	2.14	2.09
40	4.08	3.23	2.84	2.61	2.45	2.34	2.25	2.18	2.12	2.08
42	4.07	3.22	2.83	2.59	2.44	2.32	2.24	2.16	2.11	2.06
44	4.06	3.21	2.82	2.58	2.43	2.31	2.23	2.16	2.10	2.05
46	4.05	3.20	2.81	2.57	2.42	2.30	2.22	2.15	2.09	2.04
48	4.04	3.19	2.80	2.57	2.41	2.29	2.21	2.14	2.08	2.03
50	4.03	3.18	2.79	2.56	2.40	2.29	2.20	2.13	2.07	2.03
55	4.02	3.16	2.77	2.54	2.38	2.27	2.18	2.11	2.06	2.01
60	4.00	3.15	2.76	2.53	2.37	2.25	2.17	2.10	2.04	1.99
65	3.99	3.14	2.75	2.51	2.36	2.24	2.15	2.08	2.03	1.98
70	3.98	3.13	2.74	2.50	2.35	2.23	2.14	2.07	2.02	1.97
80	3.96	3.11	2.73	2.49	2.33	2.21	2.13	2.06	2.00	1.95

Degrees of freedom for the denominator (ν_2)

TABLE D.4 *F* Values (Continued)

$F_{.05;\nu_1;\nu_2}$ 95% confidence‡

		1	2	3	4	5	6	7	8	9	10
Degrees of freedom for the denominator (ν_2)	90	3.95	3.10	2.71	2.47	2.32	2.20	2.11	2.04	1.99	1.94
	100	3.94	3.09	2.70	2.46	2.31	2.19	2.10	2.03	1.97	1.93
	125	3.92	3.07	2.68	2.44	2.29	2.17	2.08	2.01	1.96	1.91
	150	3.90	3.08	2.66	2.43	2.27	2.16	2.07	2.00	1.94	1.89
	200	3.89	3.04	2.65	2.42	2.26	2.14	2.06	1.98	1.93	1.88
	300	3.87	3.03	2.63	2.40	2.24	2.13	2.04	1.97	1.91	1.86
	500	3.86	3.01	2.62	2.39	2.23	2.12	2.03	1.96	1.90	1.85
	1000	3.85	3.00	2.61	2.38	2.22	2.11	2.02	1.95	1.89	1.84
	∞	3.84	3.00	2.60	2.37	2.21	2.10	2.01	1.94	1.88	1.83

Degrees of freedom for the numerator (ν_1) (column headers 1–10 above)

$F_{.01;\nu_1;\nu_2}$ 99% confidence#

Multiply the numbers of the first row ($\nu_2 = 1$) by 10

	1	2	3	4	5	6	7	8	9	10
1	405	500	540	563	576	596	598	598	602	606
2	93.5	99.0	99.3	99.3	99.3	99.3	99.4	99.4	99.4	99.4
3	34.1	30.8	20.5	28.7	28.2	27.9	27.7	27.5	27.3	27.2
4	21.2	18.0	16.7	16.0	15.5	15.2	15.0	14.8	14.7	14.5
5	16.8	13.2	12.1	11.4	11.0	10.7	10.5	10.3	10.2	10.1
6	13.7	10.9	9.78	9.15	8.75	8.47	8.28	8.10	7.98	7.87
7	12.2	9.55	8.45	7.85	7.46	7.19	6.99	6.94	6.72	6.62
8	11.3	8.65	7.89	7.01	6.63	6.37	6.18	6.03	5.91	5.81
9	10.6	8.02	6.99	6.42	6.06	5.80	5.61	5.47	5.35	5.26
10	10.0	7.56	6.55	5.99	5.64	5.39	5.20	5.06	4.94	4.85
11	9.65	7.21	6.22	5.67	5.32	5.07	4.89	4.74	4.63	4.54
12	9.33	6.93	5.95	5.41	5.06	4.82	4.64	4.50	4.30	4.30
13	9.07	6.70	5.74	5.21	4.86	4.62	4.44	4.30	4.19	4.10
14	8.86	6.51	5.58	5.04	4.70	4.46	4.28	4.14	4.03	3.94
15	8.68	6.26	5.42	4.89	4.56	4.32	4.14	4.00	3.89	3.80
16	8.53	6.22	5.29	4.77	4.44	4.20	4.03	3.89	3.78	3.69
17	8.60	6.11	5.18	4.67	4.34	4.10	3.93	3.79	3.68	3.59
18	8.20	6.01	5.09	4.58	4.25	4.01	3.84	3.71	3.60	3.51
19	8.18	5.93	5.01	4.50	4.17	3.94	3.77	3.68	3.52	3.43
20	8.10	5.85	4.94	4.43	4.10	3.87	3.70	3.56	3.46	3.37
21	8.02	5.78	4.87	4.37	4.04	3.81	3.64	3.51	3.40	3.31
22	7.95	5.72	4.82	4.31	3.99	3.76	3.59	3.45	3.35	3.26
23	7.86	5.66	4.76	4.26	3.94	3.71	3.54	3.41	3.30	3.21
24	7.82	5.61	4.72	4.22	3.90	3.67	3.50	3.36	3.26	3.17
25	7.77	5.57	4.68	4.18	3.86	3.63	3.46	3.32	3.22	3.13
26	7.72	5.53	4.64	4.14	3.82	3.59	3.42	3.29	3.18	3.09
27	7.66	5.49	4.60	4.11	3.78	3.56	3.39	3.26	3.15	3.06
28	7.64	5.45	4.57	4.07	3.75	3.53	3.36	3.23	3.12	3.03
29	7.60	5.42	4.54	4.04	3.73	3.50	3.33	3.20	3.09	3.00
30	7.56	5.39	4.51	4.03	3.70	3.47	3.30	3.17	3.07	2.98

Degrees of freedom for the denominator (ν_2)

TABLE D.4 *F* Values (Continued)

	$F_{.01;\nu_1;\nu_2}$			99% confidence#						
	Degrees of freedom for the numerator (ν_1)									
	1	2	3	4	5	6	7	8	9	10
32	7.50	5.34	4.46	3.97	3.65	3.43	3.26	3.13	3.02	2.93
34	7.44	5.29	4.42	3.93	3.61	3.39	3.23	3.09	2.96	2.89
36	7.40	5.25	4.36	3.89	3.57	3.35	3.18	3.05	2.95	2.86
38	7.35	5.21	4.34	3.86	3.54	3.32	3.15	3.02	2.92	2.83
40	7.31	5.18	4.31	3.83	3.51	3.29	3.12	2.99	2.80	2.80
42	7.28	5.15	4.29	3.80	3.49	3.27	3.10	2.97	2.86	2.78
44	7.25	5.12	4.26	3.78	3.47	3.24	3.08	2.95	2.84	2.75
46	7.22	5.10	4.24	3.76	3.44	3.22	3.06	2.93	2.82	2.73
48	7.19	5.08	4.22	3.74	3.43	3.20	3.04	2.91	2.80	2.72
50	7.17	5.06	4.20	3.72	3.41	3.19	3.02	2.89	2.79	2.70
55	7.12	5.01	4.16	3.68	3.37	3.15	2.98	2.85	2.75	2.66
60	7.08	4.98	4.12	3.65	3.34	3.12	2.95	2.82	2.72	2.63
65	7.04	4.95	4.10	3.62	3.31	3.09	2.93	2.80	2.69	2.61
70	7.01	4.92	4.08	3.60	3.29	3.07	2.91	2.78	2.67	2.59
80	6.98	4.88	4.04	3.56	3.26	3.04	2.87	2.74	2.64	2.55
90	6.93	4.85	4.01	3.54	3.23	3.01	2.84	2.72	2.61	2.52
100	6.90	4.83	3.96	3.51	3.21	2.99	2.82	2.69	2.59	2.50
125	6.84	4.78	3.94	3.47	3.17	2.95	2.79	2.66	2.55	2.47
150	6.81	4.75	3.92	3.45	3.14	2.92	2.76	2.63	2.53	2.44
200	6.76	4.71	3.88	3.41	3.11	2.89	2.73	2.60	2.50	2.41
300	6.72	4.68	3.85	3.38	3.08	2.86	2.70	2.57	2.47	2.36
500	6.69	4.65	3.82	3.36	3.05	2.84	2.68	2.55	2.44	2.36
1000	6.66	4.63	3.80	3.34	3.04	2.82	2.66	2.53	2.43	2.34
∞	6.66	4.61	3.78	3.32	3.02	2.80	2.64	2.51	2.41	2.32

Degrees of freedom for the denominator (ν_2)

TABLE D.5 Omega Conversion Table*

p, %	db	p, %	db	p, %	db
0.0	∞	5.0	−12.787	10.0	−9.541
0.1	−29.995	5.1	−12.696	10.1	−9.493
0.2	−26.980	5.2	−12.607	10.2	−9.446
0.3	−25.215	5.3	−12.520	10.3	−9.399
0.4	−23.961	5.4	−12.434	10.4	−9.352
0.5	−22.988	5.5	−12.350	10.5	−9.305
0.6	−22.191	5.6	−12.267	10.6	−9.259
0.7	−21.518	5.7	−12.185	10.7	−9.214
0.8	−20.933	5.8	−12.105	10.8	−9.168
0.9	−20.417	5.9	−12.026	10.9	−9.124
1.0	−19.955	6.0	−11.949	11.0	−9.079
1.1	−19.537	6.1	−11.872	11.1	−9.035
1.2	−19.155	6.2	−11.797	11.2	−8.991
1.3	−18.803	6.3	−11.723	11.3	−8.947
1.4	−18.476	6.4	−11.650	11.4	−8.904
1.5	−18.172	6.5	−11.578	11.5	−8.861
1.6	−17.888	6.6	−11.507	11.6	−8.819
1.7	−17.620	6.7	−11.437	11.7	−8.777
1.8	−17.367	6.8	−11.368	11.8	−8.735
1.9	−17.128	6.9	−11.300	11.9	−8.693
2.0	−16.901	7.0	−11.233	12.0	−8.652
2.1	−16.685	7.1	−11.167	12.1	−8.611
2.2	−16.478	7.2	−11.101	12.2	−8.570
2.3	−16.281	7.3	−11.037	12.3	−8.530
2.4	−16.091	7.4	−10.973	12.4	−8.490
2.5	−15.910	7.5	−10.910	12.5	−8.450
2.6	−15.735	7.6	−10.848	12.6	−8.410
2.7	−15.566	7.7	−10.786	12.7	−8.371
2.8	−15.404	7.8	−10.725	12.8	−8.332
2.9	−15.247	7.9	−10.665	12.9	−8.293
3.0	−15.096	8.0	−10.606	13.0	−8.255
3.1	−14.949	8.1	−10.547	13.1	−8.216
3.2	−14.806	8.2	−10.489	13.2	−8.178
3.3	−14.668	8.3	−10.432	13.3	−8.141
3.4	−14.534	8.4	−10.375	13.4	−8.103
3.5	−14.404	8.5	−10.319	13.5	−8.066
3.6	−14.227	8.6	−10.263	13.6	−8.029
3.7	−14.153	8.7	−10.209	13.7	−7.992
3.8	−14.033	8.8	−10.154	13.8	−7.955
3.9	−13.916	8.9	−10.100	13.9	−7.919
4.0	−13.801	9.0	−10.047	14.0	−7.883
4.1	−13.689	9.1	−9.994	14.1	−7.847
4.2	−13.580	9.2	−9.942	14.2	−7.811
4.3	−13.473	9.3	−9.890	14.3	−7.775
4.4	−13.369	9.4	−9.839	14.4	−7.740
4.5	−13.267	9.5	−9.788	14.5	−7.705
4.6	−13.167	9.6	−9.738	14.6	−7.670
4.7	−13.069	9.7	−9.688	14.7	−7.635
4.8	−12.973	9.8	−9.639	14.8	−7.601
4.9	−12.879	9.9	−9.590	14.9	−7.566

*Genichi Taguchi and Yu-In Wu, *Off-Line Quality Control*, Central Japan Quality Control Association, Nagaya, 1979, pp. 99–102.

TABLE D.5   Omega Conversion Table (Continued)

p, %	db	p, %	db	p, %	db
15.0	−7.532	20.0	−6.020	25.0	−4.770
15.1	−7.498	20.1	−5.993	25.1	−4.747
15.2	−7.465	20.2	−5.966	25.2	−4.724
15.3	−7.431	20.3	−5.939	25.3	−4.701
15.4	−7.397	20.4	−5.912	25.4	−4.678
15.5	−7.364	20.5	−5.885	25.5	−4.655
15.6	−7.331	20.6	−5.859	25.6	−4.632
15.7	−7.298	20.7	−5.832	25.7	−4.610
15.8	−7.266	20.8	−5.806	25.8	−4.587
15.9	−7.233	20.9	−5.779	25.9	−4.564
16.0	−7.201	21.0	−5.753	26.0	−4.542
16.1	−7.168	21.1	−5.727	26.1	−4.519
16.2	−7.136	21.2	−5.701	26.2	−4.497
16.3	−7.104	21.3	−5.675	26.3	−4.474
16.4	−7.073	21.4	−5.649	26.4	−4.452
16.5	−7.041	21.5	−5.623	26.5	−4.429
16.6	−7.010	21.6	−5.598	26.6	−4.407
16.7	−6.978	21.7	−5.572	26.7	−4.385
16.8	−6.947	21.8	−5.547	26.8	−4.363
16.9	−6.916	21.9	−5.521	26.9	−4.341
17.0	−6.885	22.0	−5.496	27.0	−4.319
17.1	−6.855	22.1	−5.470	27.1	−4.297
17.2	−6.824	22.2	−5.445	27.2	−4.275
17.3	−6.794	22.3	−5.420	27.3	−4.253
17.4	−6.763	22.4	−5.395	27.4	−4.231
17.5	−6.733	22.5	−5.370	27.5	−4.209
17.6	−6.703	22.6	−5.345	27.6	−4.187
17.7	−6.673	22.7	−5.321	27.7	−4.166
17.8	−6.644	22.8	−5.296	27.8	−4.144
17.9	−6.614	22.9	−5.271	27.9	−4.122
18.0	−6.584	23.0	−5.427	28.0	−4.101
18.1	−6.555	23.1	−5.222	28.1	−4.079
18.2	−6.526	23.2	−5.198	28.2	−4.058
18.3	−6.497	23.3	−5.173	28.3	−4.036
18.4	−6.468	23.4	−5.149	28.4	−4.015
18.5	−6.439	23.5	−5.125	28.5	−3.994
18.6	−6.410	23.6	−5.101	28.6	−3.972
18.7	−6.381	23.7	−5.077	28.7	−3.951
18.8	−6.353	23.8	−5.053	28.8	−3.930
18.9	−6.325	23.9	−5.029	28.9	−3.909
19.0	−6.296	24.0	−5.005	29.0	−3.888
19.1	−6.268	24.1	−4.981	29.1	−3.867
19.2	−6.240	24.2	−4.958	29.2	−3.846
19.3	−6.212	24.3	−4.934	29.3	−3.825
19.4	−6.184	24.4	−4.910	29.4	−3.804
19.5	−6.157	24.5	−4.887	29.5	−3.783
19.6	−6.129	24.6	−4.863	29.6	−3.762
19.7	−6.101	24.7	−4.840	29.7	−3.741
19.8	−6.074	24.8	−4.817	29.8	−3.720
19.9	−6.047	24.9	−4.793	29.9	−3.699

TABLE D.5   Omega Conversion Table (Continued)

p, %	db	p, %	db	p, %	db
30.0	−3.679	35.0	−2.687	40.0	−1.760
30.1	−3.658	35.1	−2.668	40.1	−1.742
30.2	−3.637	35.2	−2.649	40.2	−1.724
30.3	−3.617	35.3	−2.630	40.3	−1.706
30.4	−3.596	35.4	−2.611	40.4	−1.688
30.5	−3.576	35.5	−2.592	40.5	−1.670
30.6	−3.555	35.6	−2.573	40.6	−1.652
30.7	−3.535	35.7	−2.554	40.7	−1.634
30.8	−3.515	35.8	−2.536	40.8	−1.616
30.9	−3.494	35.9	−2.517	40.9	−1.598
31.0	−3.474	36.0	−2.498	41.0	−1.580
31.1	−3.454	36.1	−2.479	41.1	−1.562
31.2	−3.433	36.2	−2.460	41.2	−1.544
31.3	−3.413	36.3	−2.441	41.3	−1.526
31.4	−3.393	36.4	−2.423	41.4	−1.508
31.5	−3.373	36.5	−2.404	41.5	−1.490
31.6	−3.353	36.6	−2.385	41.6	−1.472
31.7	−3.333	36.7	−2.366	41.7	−1.454
31.8	−3.313	36.8	−2.348	41.8	−1.436
31.9	−3.293	36.9	−2.329	41.9	−1.419
32.0	−3.273	37.0	−2.310	42.0	−1.401
32.1	−3.253	37.1	−2.292	42.1	−1.383
32.2	−3.233	37.2	−2.273	42.2	−1.365
32.3	−3.213	37.3	−2.255	42.3	−1.347
32.4	−3.193	37.4	−2.236	42.4	−1.330
32.5	−3.173	37.5	−2.217	42.5	−1.312
32.6	−3.153	37.6	−2.199	42.6	−1.294
32.7	−3.134	37.7	−2.180	42.7	−1.276
32.8	−3.114	37.8	−2.162	42.8	−1.259
32.9	−3.094	37.9	−2.144	42.9	−1.241
33.0	−3.075	38.0	−2.125	43.0	−1.223
33.1	−3.055	38.1	−2.107	43.1	−1.205
33.2	−3.035	38.2	−2.088	43.2	−1.188
33.3	−3.016	38.3	−2.070	43.3	−1.170
33.4	−2.996	38.4	−2.051	43.4	−1.152
33.5	−2.977	38.5	−2.033	43.5	−1.135
33.6	−2.957	38.6	−2.015	43.6	−1.117
33.7	−2.938	38.7	−1.996	43.7	−1.009
33.8	−2.918	38.8	−1.978	43.8	−1.082
33.9	−2.899	38.9	−1.960	43.9	−1.064
34.0	−2.880	39.0	−1.942	44.0	−1.046
34.1	−2.860	39.1	−1.923	44.1	−1.029
34.2	−2.841	39.2	−1.905	44.2	−1.011
34.3	−2.822	39.3	−1.887	44.3	−0.994
34.4	−2.802	39.4	−1.869	44.4	−0.976
34.5	−2.783	39.5	−1.851	44.5	−0.958
34.6	−2.764	39.6	−1.832	44.6	−0.941
34.7	−2.745	39.7	−1.814	44.7	−0.923
34.8	−2.726	39.8	−1.796	44.8	−0.906
34.9	−2.707	39.9	−1.778	44.9	−0.888

TABLE D.5 Omega Conversion Table (Continued)

p, %	db	p, %	db	p, %	db
45.0	−0.871	50.0	0.000	55.0	0.872
45.1	−0.853	50.1	0.017	55.1	0.889
45.2	−0.835	50.2	0.035	55.2	0.907
45.3	−0.818	50.3	0.052	55.3	0.924
45.4	−0.800	50.4	0.069	55.4	0.942
45.5	−0.783	50.5	0.087	55.5	0.959
45.6	−0.765	50.6	0.104	55.6	0.977
45.7	−0.748	50.7	0.122	55.7	0.995
45.8	−0.730	50.8	0.139	55.8	1.012
45.9	−0.713	50.9	0.156	55.9	1.030
46.0	−0.695	51.0	0.174	56.0	1.047
46.1	−0.678	51.1	0.191	56.1	1.065
46.2	−0.660	51.2	0.209	56.2	1.083
46.3	−0.643	51.3	0.226	56.3	1.100
46.4	−0.625	51.4	0.243	56.4	1.118
46.5	−0.608	51.5	0.261	56.5	1.136
46.6	−0.591	51.6	0.278	56.6	1.153
46.7	−0.573	51.7	0.295	56.7	1.171
46.8	−0.556	51.8	0.313	56.8	1.189
46.9	−0.538	51.9	0.330	56.9	1.206
47.0	−0.521	52.0	0.348	57.0	1.224
47.1	−0.503	52.1	0.365	57.1	1.242
47.2	−0.486	52.2	0.382	57.2	1.260
47.3	−0.468	52.3	0.400	57.3	1.277
47.4	−0.451	52.4	0.417	57.4	1.295
47.5	−0.434	52.5	0.435	57.5	1.313
47.6	−0.416	52.6	0.452	57.6	1.331
47.7	−0.399	52.7	0.469	57.7	1.348
47.8	−0.381	52.8	0.487	57.8	1.366
47.9	−0.364	52.9	0.504	57.9	1.384
48.0	−0.347	53.0	0.522	58.0	1.402
48.1	−0.329	53.1	0.539	58.1	1.420
48.2	−0.312	53.2	0.557	58.2	1.437
48.3	−0.294	53.3	0.574	58.3	1.455
48.4	−0.277	53.4	0.592	58.4	1.473
48.5	−0.260	53.5	0.609	58.5	1.491
48.6	−0.242	53.6	0.626	58.6	1.509
48.7	−0.225	53.7	0.644	58.7	1.527
48.8	−0.208	53.8	0.661	58.8	1.545
48.9	−0.190	53.9	0.679	58.9	1.563
49.0	−0.173	54.0	0.696	59.0	1.581
49.1	−0.155	54.1	0.714	59.1	1.599
49.2	−0.138	54.2	0.731	59.2	1.617
49.3	−0.121	54.3	0.749	59.3	1.635
49.4	−0.103	54.4	0.679	59.4	1.653
49.5	−0.086	54.5	0.784	59.5	1.671
49.6	−0.068	54.6	0.801	59.6	1.689
49.7	−0.051	54.7	0.819	59.7	1.707
49.8	−0.034	54.8	0.836	59.8	1.725
49.9	−0.016	54.9	0.854	59.9	1.743

− Values | + Values

TABLE D.5   Omega Conversion Table (Continued)

p, %	db	p, %	db	p, %	db
60.0	1.761	65.0	2.688	70.0	3.680
60.1	1.779	65.1	2.708	70.1	3.700
60.2	1.797	65.2	2.727	70.2	3.721
60.3	1.815	65.3	2.746	70.3	3.742
60.4	1.833	65.4	2.765	70.4	3.763
60.5	1.852	65.5	2.784	70.5	3.784
60.6	1.870	65.6	2.803	70.6	3.805
60.7	1.888	65.7	2.823	70.7	3.826
60.8	1.906	65.8	2.842	70.8	3.847
60.9	1.924	65.9	2.861	70.9	3.868
61.0	1.943	66.0	2.881	71.0	3.889
61.1	1.961	66.1	2.900	71.1	3.910
61.2	1.979	66.2	2.919	71.2	3.931
61.3	1.997	66.3	2.939	71.3	3.952
61.4	2.016	66.4	2.958	71.4	3.973
61.5	2.034	66.5	2.978	71.5	3.995
61.6	2.052	66.6	2.997	71.6	4.016
61.7	2.071	66.7	3.017	71.7	4.037
61.8	2.089	66.8	3.036	71.8	4.059
61.9	2.108	66.9	3.056	71.9	4.080
62.0	2.126	67.0	3.076	72.0	4.102
62.1	2.145	67.1	3.095	72.1	4.123
62.2	2.163	67.2	3.115	72.2	4.145
62.3	2.181	67.3	3.135	72.3	4.167
62.4	2.200	67.4	3.154	72.4	4.188
62.5	2.218	67.5	3.174	72.5	4.210
62.6	2.237	67.6	3.194	72.6	4.232
62.7	2.256	67.7	3.214	72.7	4.254
62.8	2.274	67.8	3.234	72.8	4.276
62.9	2.293	67.9	3.254	72.9	4.298
63.0	2.311	68.0	3.274	73.0	4.320
63.1	2.330	68.1	3.294	73.1	4.342
63.2	2.349	68.2	3.314	73.2	4.364
63.3	2.367	68.3	3.334	73.3	4.386
63.4	2.386	68.4	3.354	73.4	4.408
63.5	2.405	68.5	3.374	73.5	4.430
63.6	2.424	68.6	3.394	73.6	4.453
63.7	2.442	68.7	3.414	73.7	4.475
63.8	2.461	68.8	3.434	73.8	4.498
63.9	2.480	68.9	3.455	73.9	4.520
64.0	2.499	69.0	3.475	74.0	4.543
64.1	2.518	69.1	3.495	74.1	4.565
64.2	2.537	69.2	3.516	74.2	4.588
64.3	2.555	69.3	3.536	74.3	4.611
64.4	2.574	69.4	3.556	74.4	4.633
64.5	2.593	69.5	3.577	74.5	4.656
64.6	2.612	69.6	3.597	74.6	4.679
64.7	2.631	69.7	3.618	74.7	4.702
64.8	2.650	69.8	3.638	74.8	4.725
64.9	2.669	69.9	3.659	74.9	4.748

TABLE D.5 Omega Conversion Table (Continued)

p, %	db	p, %	db	p, %	db
75.0	4.771	80.0	6.021	85.0	7.533
75.1	4.794	80.1	6.048	85.1	7.567
75.2	4.818	80.2	6.075	85.2	7.602
75.3	4.841	80.3	6.102	85.3	7.686
75.4	4.864	80.4	6.130	85.4	7.671
75.5	4.888	80.5	6.158	85.5	7.706
75.6	4.911	80.6	6.185	85.6	7.741
75.7	4.935	80.7	6.213	85.7	7.776
75.8	4.959	80.8	6.241	85.8	7.812
75.9	4.982	80.9	6.269	85.9	7.848
76.0	5.006	81.0	6.297	86.0	7.884
76.1	5.030	81.1	6.326	86.1	7.920
76.2	5.054	81.2	6.354	86.2	7.956
76.3	5.078	81.3	6.382	86.3	7.993
76.4	5.102	81.4	6.411	86.4	8.080
76.5	5.126	81.5	6.440	86.5	8.067
76.6	5.150	81.6	6.469	86.6	8.104
76.7	5.174	81.7	6.498	86.7	8.142
76.8	5.199	81.8	6.527	86.8	8.179
76.9	5.223	81.9	6.556	86.9	8.217
77.0	5.248	82.0	6.585	87.0	8.256
77.1	5.272	82.1	6.615	87.1	8.294
77.2	5.297	82.2	6.645	87.2	8.333
77.3	5.322	82.3	6.674	87.3	8.372
77.4	5.346	82.4	6.704	87.4	8.411
77.5	5.371	82.5	6.734	87.5	8.451
77.6	5.396	82.6	6.764	87.6	8.491
77.7	5.421	82.7	6.795	87.7	8.531
77.8	5.446	82.8	6.825	87.8	8.571
77.9	5.471	82.9	6.856	87.9	8.612
78.0	5.497	83.0	6.886	88.0	8.653
78.1	5.522	83.1	6.917	88.1	8.694
78.2	5.548	83.2	6.948	88.2	8.786
78.3	5.573	83.3	6.979	88.3	8.778
78.4	5.599	83.4	7.011	88.4	8.820
78.5	5.624	83.5	7.042	88.5	8.862
78.6	5.650	83.6	7.074	88.6	8.905
78.7	5.676	83.7	7.105	88.7	8.948
78.8	5.702	83.8	7.137	88.8	8.992
78.9	5.728	83.9	7.169	88.9	9.036
79.0	5.754	84.0	7.202	89.0	9.080
79.1	5.780	84.1	7.234	89.1	9.125
79.2	5.807	84.2	7.267	89.2	9.169
79.3	5.833	84.3	7.299	89.3	9.215
79.4	5.860	84.4	7.332	89.4	9.260
79.5	5.886	84.5	7.365	89.5	9.306
79.6	5.913	84.6	7.398	89.6	9.353
79.7	5.940	84.7	7.432	89.7	9.400
79.8	5.967	84.8	7.466	89.8	9.447
79.9	5.994	84.9	7.499	89.9	9.494

TABLE D.5 Omega Conversion Table (Continued)

p, %	db	p, %	db	p, %	db
90.0	9.542	94.0	11.950	98.0	16.902
90.1	9.591	94.1	12.027	98.1	17.129
90.2	9.640	94.2	12.106	98.2	17.368
90.3	9.689	94.3	12.186	98.3	17.621
90.4	9.739	94.4	12.268	98.4	17.889
90.5	9.789	94.5	12.351	98.5	18.173
90.6	9.840	94.6	12.435	98.6	18.447
90.7	9.891	94.7	12.521	98.7	18.804
90.8	9.943	94.8	12.608	98.8	19.156
90.9	9.995	94.9	12.697	98.9	19.538
91.0	10.048	95.0	12.783	99.0	19.956
91.1	10.111	95.1	12.880	99.1	20.418
91.2	10.155	95.2	12.974	99.2	20.934
91.3	10.210	95.3	13.070	99.3	21.519
91.4	10.264	95.4	13.168	99.4	22.192
91.5	10.320	95.5	13.268	99.5	22.989
91.6	10.376	95.6	13.370	99.6	23.962
91.7	10.433	95.7	13.474	99.7	25.216
91.8	10.490	95.8	13.581	99.8	26.981
91.9	10.548	95.9	13.690	99.9	29.996
92.0	10.607	96.0	13.802	100.0	∞
92.1	10.666	96.1	13.917		
92.2	10.726	96.2	14.034		
92.3	10.787	96.3	14.154		
92.4	10.840	96.4	14.278		
92.5	10.911	96.5	14.405		
92.6	10.974	96.6	14.535		
92.7	11.038	96.7	14.669		
92.8	11.102	96.8	14.807		
92.9	11.168	96.9	14.950		
93.0	11.234	97.0	15.097		
93.1	11.301	97.1	15.248		
93.2	11.369	97.2	15.405		
93.3	11.438	97.3	15.567		
93.4	11.508	97.4	15.736		
93.5	11.579	97.5	15.911		
93.6	11.651	97.6	16.092		
93.7	11.724	97.7	16.282		
93.8	11.798	97.8	16.479		
93.9	11.873	97.9	16.686		

$$\Omega(db) = 10 \log \left[\frac{p/100}{1 - p/100} \right]$$

Proofs

Proof 1 (see page 12)

$S^2_{y;m} = S^2_{y;\bar{y}} + (\bar{y} - m)^2$

variance of y with baseline of m = variance of y with baseline of $\bar{y}$ + square of the difference of average and m

$$\frac{\sum\limits_{i=1}^{n} (y_i - m)^2}{n} = \frac{\sum\limits_{i=1}^{n} (y_i - \bar{y})^2}{n} + (\bar{y} - m)^2$$

$$\sum (y_i - m)^2 = \sum (y_i - \bar{y})^2 + n(\bar{y} - m)^2$$

$$\sum (y_i^2 - 2my_i + m^2) = \sum (y_i^2 - 2\bar{y}y_i + \bar{y}^2) + n(\bar{y}^2 - 2m\bar{y} + m^2)$$

$$\sum y_i^2 - 2m \sum y_i + nm^2 = \sum y_i^2 - 2\bar{y} \sum y_i + n\bar{y}^2 + n\bar{y}^2 - 2mn\bar{y} + nm^2$$

$$-2m \sum y_i = -2\bar{y} \sum y_i + 2n\bar{y}^2 - 2mn\bar{y}$$

$$-m \sum y_i = -\bar{y} \sum y_i + n\bar{y}^2 - mn\bar{y}$$

Given

$$\bar{y} = \frac{\sum y_i}{n}$$

$$-m \sum y_i = \frac{-\sum y_i \left(\sum y_i\right)}{n} + n\left(\frac{\sum y_i}{n}\right)^2 - mn\left(\frac{\sum y_i}{n}\right)$$

$$0 = -\frac{\left(\sum y_i\right)^2}{n} + \frac{\left(\sum y_i\right)^2}{n}$$

$$0 = 0$$

Proof 2 (see page 34)

$$SS_A = n(\overline{A}_1 - \overline{T})^2 + n(\overline{A}_2 - \overline{T})^2$$

Given: $n + n = N = 2n$

$$SS_A = n[(\overline{A}_1 - \overline{T})^2 + (\overline{A}_2 - \overline{T})^2]$$

$$= n(\overline{A}_1^2 - 2\overline{A}_1\overline{T} + \overline{T}^2 + \overline{A}_2^2 - 2\overline{A}_2\overline{T} + \overline{T}^2)$$

$$= n[\overline{A}_1^2 + \overline{A}_2^2 - 2\overline{T}(\overline{A}_1 + \overline{A}_2) + 2\overline{T}^2]$$

but $2\overline{T} = \overline{A}_1 + \overline{A}_2$

$$SS_A = n[\overline{A}_1^2 + \overline{A}_2^2 - 2\overline{T}(2\overline{T}) + 2\overline{T}^2]$$

$$= n(\overline{A}_1^2 + \overline{A}_2^2 - 4\overline{T}^2 + 2\overline{T}^2)$$

$$= n(\overline{A}_1^2 + \overline{A}_2^2 - 2\overline{T}^2)$$

$$= n\overline{A}_1^2 + n\overline{A}_2^2 - 2n\overline{T}^2$$

but $\overline{A}_1 = A_1/n$; $\overline{A}_2 = A_2/n$; and $\overline{T} = T/2n$

$$SS_A = n\left(\frac{A_1}{n}\right)^2 + n\left(\frac{A_2}{n}\right)^2 - 2n\left(\frac{T}{2n}\right)^2$$

$$= \frac{A_1^2}{n} + \frac{A_2^2}{n} - \frac{T^2}{2n}$$

$$= \frac{A_1^2}{n} + \frac{A_2^2}{n} - \frac{T^2}{N}$$

This is the proof of the general formula for sum of squares using a two-level example.

but $T = A_1 + A_2$

$$SS_A = \frac{A_1^2}{n} + \frac{A_2^2}{n} - \frac{(A_1 + A_2)^2}{N}$$

$$= \frac{(A_1^2 + A_2^2)}{n} - \frac{(A_1^2 + 2A_1A_2 + A_2^2)}{2n}$$

$$= \frac{2(A_1^2 + A_2^2)}{2n} - \frac{(A_1^2 + 2A_1A_2 + A_2^2)}{2n}$$

$$= \frac{(2A_1^2 + 2A_2^2 - A_1^2 - 2A_1A_2 - A_2^2)}{2n}$$

$$= \frac{(A_1^2 + A_2^2 - 2A_1A_2)}{2n}$$

$$= \frac{(A_1^2 - 2A_1A_2 + A_2^2)}{N}$$

$$= \frac{(A_1 - A_2)^2}{N}$$

This is the proof of the specific formula for sum of squares.

Proof 3 (see page 38)

$$\sum_{i=1}^{N} (y_i - \overline{T})^2 = \left[\sum_{i=1}^{N} y_i^2 \right] - \frac{T^2}{N}$$

$$\sum (y_i^2 - 2\overline{T}y_i + \overline{T}^2) = \left[\sum_{i=1}^{N} y_i^2 \right] - \frac{T^2}{N}$$

$$\sum y_i^2 - \sum 2\overline{T}y_i + \sum \overline{T}^2 = \left[\sum_{i=1}^{N} y_i^2 \right] - \frac{T^2}{N}$$

$$\sum y_i^2 - 2\overline{T}\sum y_i + N\overline{T}^2 = \left[\sum_{i=1}^{N} y_i^2 \right] - \frac{T^2}{N}$$

Given $T = \sum y_i$ and $\overline{T} = T/N$

$$\left[\sum y_i^2 \right] - 2\left(\frac{T}{N} \right)T + N\overline{T}^2 = \left[\sum_{i=1}^{N} y_i^2 \right] - \frac{T^2}{N}$$

$$\left[\sum y_i^2 \right] - \frac{2T^2}{N} + \frac{T^2}{N} = \left[\sum_{i=1}^{N} y_i^2 \right] - \frac{T^2}{N}$$

$$\left[\sum y_i^2 \right] - \frac{T^2}{N} = \left[\sum y_i^2 \right] - \frac{T^2}{N}$$

Definitions and Symbols

Definitions

Alpha (α) error The probability that the null hypothesis will be rejected when it is in fact true; the chance of saying that a mean of a population is different when in fact it is equal to another population mean; type I error; the producer's risk of specifying a factor condition which is not really helpful.

Alternative hypothesis (H_a) The assumption that the null hypothesis is not true; there is some inequality existing between the values being compared; the alternative hypothesis can be proven true at some confidence level which is always $(1 - \alpha)$.

Analysis of variance (ANOVA) A procedure to isolate one source of variation from another; a method to decompose the total variation present into accountable sources; a method to make a statistically based decision as to the causes of variation in an experiment.

Average See *Mean of sample.*

Beta (β) error The probability that the null hypothesis will be accepted when it is in fact not true; the chance of saying there is no difference of two population means when in fact a difference does exist; type II error; the consumer's risk of a producer missing a helpful factor.

Binomial The distribution of a discrete variable or attribute where there are two possible outcomes, the probability of which remains constant, such as tossing coins, passing or failing a test, fitting a go–no go gauge, having a defect present or not present, etc.

Blocking   A technique used in factorial experimentation to reduce the effect of unwanted variation by restricting the design of the experiment. A block of an experiment is a portion of the total experimental material that is expected to be more uniform than the whole because of more controlled conditions than if randomized. Treating a known source of variation as a factor.

Central limit theorem   Statistical theorem which states that sample means will tend to be normally distributed around the population mean with less variance than the individuals.

Component search   A procedure that is a derivative of factorial experiments in which components are interchanged between what is a good-performing unit and a poor-performing unit; in typical industrial situations only the poor unit is retested with other parts of unknown quality in an attempt to improve its performance.

Confidence interval (band)   The region enclosed by two limits of the performance characteristic which will contain the predicted mean at the stated confidence.

Confidence level   The value of one minus the risk involved $(1 - \alpha)$ or $(1 - \beta)$; stacking the odds in your favor that a correct decision will be made.

Confounded   To be unable to decompose certain sources of variation from one another in a particular experiment; a mixture of factorial effects within a particular column of an orthogonal array which are mathematically impossible to separate.

Continuous random variable   A variable which can randomly attain any value within some interval.

Control factor   A factor which can be set by a manufacturer and cannot be directly affected by the customer using the product or process.

Decomposition (of variance)   Taking the total amount of variation observed in an experiment and breaking it down into portions and assigning responsibility for that part of the variation.

Degrees of freedom (v)   The number of independent measurements available to estimate pieces of information; the number of independent (fair) comparisons that may be made within a set of data.

Design of experiments   See *Factorial experiment* and *Fractional factorial experiment*.

Discrete random variable   A variable which can randomly attain only discrete values within some interval.

Experiment   An operation under controlled conditions to determine an unknown effect; to illustrate or verify a known law; a test to establish a hypothesis.

Experimental error   Residual error; the variation observed when products are tested under "identical" conditions, a portion of which is instrumentation (measurement) repeatability.

Factorial experiment   An experiment which extracts information about several design factors more efficiently than a traditional single-factor experiment.

Fractional factorial experiment   A factorial experiment where only part of the treatment conditions are tested to more efficiently identify important factors; a Taguchi matrix is one example.

Histogram   A graphical representation of the sample frequency distribution where the width of the bar represents some interval of a variable and the height represents the frequency of occurrence of that value.

Inner noise   Functionally related noise factors which affect variation within a product (wear, fade, hardening, etc.).

Interaction ($A \times B$)   The synergistic effect of two or more factors in a factorial experiment. The effect of one factor depends on another factor.

Least square line   A line fitted to a series of test observations such that the sum of the squares of the deviations of the test observations from the line is a minimum; the best-fitting line to test data.

Levels   The settings of various factors in a factorial experiment; high and low values of pressure, temperature, etc.; may also have multiple levels of a factor; k equals the number of levels of a factor.

Linear graph   A tool used in assigning factors and interactions to a Taguchi orthogonal matrix for the purpose of experimentation (see *Triangular tables*).

Loss function   A continuous cost function which measures the cost impact of the variability of a product.

Main effect   The effect of a factor acting on its own from one level to another to change the outcome in a factorial experiment.

Mean (population) (μ)   The expected value of a population.

Mean (sample) ($\bar{x}, \bar{y}, \bar{z}$, etc.)   The summation of test observations divided by the total number of observations; average.

Nested factor   A factor which can be varied only over a portion of the entire experiment due to design or process limitations.

Noise factor   A factor that disturbs the function of a product or process; factor which may be controlled during an experiment but in the customer's typical use may not be controlled by the manufacturer; a factor that the manufacturer wishes not to control.

Null hypothesis (H_o)   Assumption that there is no inequality existing between the means of the values being compared.

Observation   The result (data) obtained when an experiment is run at particular conditions; also response.

Orthogonal matrix (array)   A fractional factorial matrix which assures a balanced, fair comparison of levels of any factor or interaction of factors; all columns can be evaluated independently of one another.

Outer noise   Environmentally related noise factors which affect variation within a product (temperature, humidity, operators, etc.).

Parameter design   A method utilizing the appropriate level of a control factor (design parameter) to reduce the sensitivity to outer, inner, and product noises.

Polynomial decomposition   Using equations to decompose the observed variation into the portions attributable to linear, quadratic, cubic, etc., functions of the factor being evaluated.

Pooling (degrees of freedom)   In an analysis of variance the combining of the sum of squares and the degrees of freedom of those factors and/or error estimates that are statistically insignificant to obtain a better estimate of experimental error.

Population   A group of similar items from which a sample is drawn for test purposes; a group of parts about which a decision is to be made.

Precision   The ability of a device to perform a measurement repeatedly with some defined variance.

Random variable   A variable which can assume any value from a set of possible values.

Repetition   Multiple test observations within a trial in an orthogonal matrix test.

Replication   Multiple test observations trial to trial in an orthogonal matrix test.

Residual error   See *Experimental error.*

Response   See *Observation.*

Sample   A random selection of items from a lot or population to evaluate the characteristics of that population; the sample must be representative of the population to be of any value.

Signal-to-noise ratio   A measure of the amount of observed variation present relative to the observed average of the data.

Standard deviation (population) (σ)   Square root of the variance of a population; measure of the scatter of the population around a mean.

Standard deviation (sample) (S)   Square root of the variance of a sample.

Sum of squares (SS)   The summation of the squared deviations relative to zero, to level means, or the grand mean of an experiment.

Tolerance design   To reduce variation of a performance characteristic by tightening tolerances on factors relating to performance; a method used when the parameter design efforts have proved to be insufficient in reducing variation.

Triangular tables   A tool used to locate the interacting columns in a Taguchi orthogonal matrix for the purpose of experimentation.

Variance (population) (σ^2)   Expected value of the square of the difference between the random variable and the mean of the population.

Variance (sample) (V or S^2)   Sum of the squares of the difference of each observation and the mean divided by the sample size minus one (for an unbiased estimate of variance); for a biased estimate of variance divide by sample size.

Symbols

A	factor of interest
A_1	sum of observations under condition A_1
$\overline{A}_1$	average of observations under condition A_1
α	alpha
β	beta
C	confidence level
F	random variable of F distribution; ratio of variances
H_a	alternative hypothesis
H_o	null hypothesis
HB	higher is better
k	the number of levels in factorial exp.; also the cost factor in the loss function
k_A	number of levels for factor A
L	loss per unit in the loss function
LB	lower is better
LSL	lower specification limit
m	nominal or target value of a characteristic
μ	population mean
ν	degrees of freedom
ν_A	degrees of freedom for factor A; equals $k_A - 1$
N	total number of observations or samples
n_{A_1}	sample size under conditions A_1
NB	nominal is best
σ	population standard deviation
σ^2	population variance
S	sample standard deviation
S^2	sample variance
SS	sum of squares
SS_T	sum of squares due to total variation
SS_m	sum of squares due to mean
SS_A	sum of squares due to factor A
SS_e	sum of squares due to error
S/N	signal-to-noise ratio
T	sum of all observations
$\overline{T}$	average of all observations

USL upper specification limit

V variance

W class weight in accumulation analysis

y individual observation or data point

y_i ith observation

$\bar{y}$ sample mean; average observation

Example Experiments

This appendix contains several examples of both designing experiments and analyzing and interpreting actual results. The first example is a simple parameter design example; the next four examples are of designing experiments; and the last three examples are of design, analysis, and interpretation. As much as possible, the examples follow the experimental procedure outlined in Sec. 8-12. The experimental factors have been renamed to protect proprietary information.

Again, the emphasis is on the experimental design stage, since analysis and interpretation is very easy with a properly structured set of tests. How to test a product or process is generally known, and with the additional knowledge of a good experimental strategy the results can be quite lucrative.

Example G.1   Engine Oil Fill Tube Cap

A contemporary engine oil fill tube cap, shown in Fig. G-1, must seal the tube against pressures generated in an engine crankcase yet be easy to install or remove. To effect a seal, the force to move the cap must be higher than the force created by crankcase pressure but be lower than a force that human fingers can provide for removal. The design of the cap provides installation and removal force by deflecting the rubber cap lip over the ridge inside the fill tube. The force required to remove or install is a function of the cap stiffness and the amount of deflection as shown by the graphs in Fig. G-2. The equation for the force is

$$\text{Force} = \text{constant} \times \text{deflection} \times \text{stiffness}$$

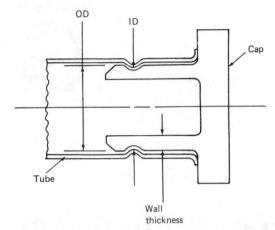

Figure G-1 Engine oil fill tube and cap.

The constant is a function of the rubber properties and the stiffness a function of the design configuration.

The stiffness of the cap can be increased by increasing the wall thickness of the cap, and the deflection can be increased by increasing the difference between the OD and ID dimensions indicated in Fig. G-1. By proper selection of the aforementioned dimensions, two stiffness curves can be generated. A parameter design approach, however, would show a preference for the low stiffness design because of the variation of the deflection from one assembly to another. The variance in the deflection is defined by the equation

$$\sigma^2_{\text{deflection}} = \sigma^2_{\text{OD}} + \sigma^2_{\text{ID}}$$

Variance is always additive in an assembly and is transmitted through to the variance in force by virtue of the stiffness relationship. The low-stiffness curve, with the same variation in deflection, transmits less variation in force and is,

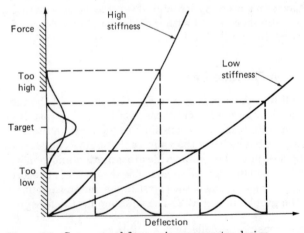

Figure G-2 Cap removal force using parameter design.

therefore, a higher-quality design, one that is more likely to meet force specifications. The loss function could be used to substantiate any cost penalties that might exist with the low-stiffness design, which would be a tolerance design approach.

Utilizing the lower-stiffness curve is similar to using the class II factors discussed in Sec. 8-5 of parameter design.

Example G.2   Windshield Washer Spray Nozzle

A windshield washer system must spray a pattern of solvent onto the glass that allows good cleaning of the windshield and is little affected by vehicle velocity. The vehicle velocity is a noise factor and the nozzle design, number of nozzle orifices, and position(s) of nozzles are control factors. The pattern tends to move (deflect) downward as velocity increases.

Problem.   To improve the spray pattern and reduce the velocity effect on the windshield washer nozzles.

Objective.   For a given vehicle design, develop a superior spray pattern and eliminate the velocity effect.

Measurement system.   The spray pattern is classified by a numerical score (1 = poor, 2 = good, 3 = superior) and the deflection can be assessed by the difference in spray position when velocity increases from zero to 100 km/h (62 mph); a grid is marked on the inside of the windshield to serve as pattern size assessment and to determine deflection.

Factors and levels.   The factor for position of the nozzles was broken into two factors; the fore-aft position and the side-to-side position. These two factors were to be evaluated over three levels to adequately cover the areas available for mounting nozzles (see Fig. G-3). The rearmost location is more desirable from a complexity viewpoint, but the hood positions may be more desirable from a pattern or deflection viewpoint. Two other factors, number of nozzle orifices and orifice diameter, are only evaluated at two levels. Table G-1 shows the factors and levels.

Assignment to an array.   Since two of the factors are three-level and two others two-level, the total degrees of freedom is equal to 6 ($v_T = 6$) and should fit into an L9 experimental array. The two-level factors will have to be dummy-treated within this type of array as shown in Table G-2.

Conducting the experiment.   The trials are completely randomized since the nozzle is movable (magnetic mount) and the nozzle heads interchangeable. The tests are conducted by adjusting nozzle angles to hit a certain spot on the windshield at zero velocity. The quality of the pattern is then observed, the vehicle accelerated to 100 km/h, and the quality of the pattern observed once again along with the change in impact point. When the data is collected, the analysis of the pattern quality and deflection would take place separately.

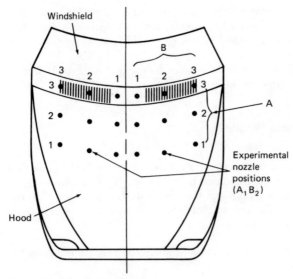

Figure G-3 Nozzle positions.

TABLE G-1 Windshield Washer Spray Factors and Levels

		Levels	
Factors	1	2	3
A Nozzle position (fore-aft)	1	2	3
B Nozzle position (side-to-side)	1	2	3
C Number of orifices	1	2	. . .
D Orifice diameter	.75 mm	1.00 mm	. . .

TABLE G-2 Windshield Washer Spray Factor Assignment to an L9 Orthogonal Array

	Factors			
	A	B	C	D
	Column no.			
Trial no.	1	2	3	4
1	1	1	1	1
2	1	2	2	2
3	1	3	1'	2'
4	2	1	2	2'
5	2	2	1'	1
6	2	3	1	2
7	3	1	1'	2
8	3	2	1	2'
9	3	3	2	1

Example G.3   Connecting Rod Bolt Torque

Engine connecting rods are typically joined at the crankshaft end by bolts on either side of the bearing cap, as shown in Fig. G-4. Consistency of the retention torque is desirable to prevent loose bearing caps and subsequent spun bearings, which will cause catastrophic engine failure. Too high a nominal bolt torque could distort the bearing journal, which would be detrimental to bearing capacity, however.

Problem.   To develop a consistent means of initially obtaining and retaining a nominal bolt clamp load. To reduce distortion of the bearing journal under bolt loads.

Objective.   To develop a design nominal clamp load, 500 lb, with a consistent bolt torque, 70 lb·ft, and minimum distortion of the journal.

Measurement system.   The clamp load is estimated by measuring bolt stretch under load. The load versus deflection relationship must be known for the bolt design. The out-of-round condition is measured by either taking several measurements before and after torque is applied with an indicator gauge or on a special machine for roundness measurements.

Factor selection and levels.   Table G-3 summarizes the control and noise factors chosen in this experiment. The torque apply method, the number of times torque is applied, the initial torque, the thread lubrication, fit of the bolt, and seat surface finish were all selected as control factors and evaluated at two levels. Noise factors to the assembly process and bolt retention were nut surface finish and bolt hardness, which were also evaluated at two levels. In this situation, the levels were selected so that the first level should provide the most consistent load conditions; the engineers had wanted to specifically test a combination of treat-

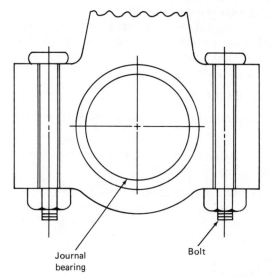

Journal
bearing

Bolt

Figure G-4 Connecting rod.

TABLE G-3 Connecting Rod Factors And Levels

		Levels	
Factors		1	2
A	Torque apply method*	Air wrench	Hand wrench
B	No. of times tightened*	3	1
C	Initial torque*	100 N·m	80 N·m
D	Thread lubrication*	Oiled	As supplied
E	Bolt fit*	Loose	Tight
F	Seat surface finish*	2 μm	5 μm
V	Nut surface finish†	2 μm	5 μm
W	Bolt hardness†	38 R_c	32 R_c

*Control factors requiring an L8 OA.
†Noise factors requiring an L4 OA.

ment conditions. Since the first trial is all the first level of factors, the combination would be assured.

Assignment of factors. The control factors were assigned to an L8 OA as the inner array and the noise factors assigned to an L4 OA as the outer array. Since one column is left over in the inner array, an interaction column was assigned for the experiment as shown in Table G-4.

Conducting the experiment. Each combination of parts is assembled under the specified conditions for the trial and the amount of bolt stretch determined, which is then translated into clamp load. Each connecting rod has two bolts. The out of roundness is determined for each connecting rod after torque is applied. The data arrangement is also shown in Table G-4.

Analyzing the data. The raw data can be analyzed to determine which factors influence the average clamp load and out of roundness. The eight repetitions of load and four repetitions of out of roundness can be transformed to one signal-to-noise ratio for nominal is best and lower is better, respectively, and analyzed to determine which factors influence (reduce) variation.

Example G.4 Automobile Tire Experiment

Problem. To reduce wear and improve traction of automobile tires on various automobiles.

Objective. To reduce wear to less than .060 in after a tire wear test and to maximize the spin-out speed in a turning circle test.

Measurement system. The amount of tread loss (after a certain amount of time on a duty cycle) determined with a depth micrometer and a fifth-wheel velocity indicated when spin-out occurs on the turning circle (the highest velocity obtained while still negotiating a specified turning radius).

Factor selection. Two-level tire control factors A, B, C, D, E, and F; four-level factors T (type of tire), and P (position of tire on car).

TABLE G-4 Connecting Rod Factor Assignment

	A	B	A×B	C	D	E	F	V W	Load				Out of round				Signal-to-noise ratio	
									1 1 1	2 2 1	2 1 2	1 2 2	1 1 1	2 2 1	2 1 2	1 2 2	L	OOR
1	1	1	1	1	1	1	1		× ×	× ×	× ×	× ×	o	o	o	o	#	*
2	1	1	1	2	2	2	2		× ×	× ×	× ×	× ×	o	o	o	o	#	*
3	1	2	2	1	1	2	2		× ×	× ×	× ×	× ×	o	o	o	o	#	*
4	1	2	2	2	2	1	1		× ×	× ×	× ×	× ×	o	o	o	o	#	*
5	2	1	2	1	2	1	2		× ×	× ×	× ×	× ×	o	o	o	o	#	*
6	2	1	2	2	1	2	1		× ×	× ×	× ×	× ×	o	o	o	o	#	*
7	2	2	1	1	2	2	1		× ×	× ×	× ×	× ×	o	o	o	o	#	*
8	2	2	1	2	1	1	2		× ×	× ×	× ×	× ×	o	o	o	o	#	*

× = load data
o = out of roundness data
# = load S/N data
* = out-of-roundness S/N data

Assignment of factors.   The factors require a total of 12 degrees of freedom, 6 degrees of freedom for the two-level factors and 3 degrees of freedom for each of the four-level factors, so an L16 OA is required at a minimum. The required linear graph is shown in Fig. G-5 and the L16 standard linear graph in Fig. G-6. One option of merged columns is 1, 2, and 3 in addition to 7, 8, and 15. The experimental arrangement is shown in Table G-5.

Example G.5   Engine Piston Scuffing

Problem.   Piston scuffing (abrasion, scoring) occurs during the final engine assembly test after initial engine build.

Objective.   To eliminate or minimize piston scuffing.

Measurement system.   The final assembly test is part of the measurement system, and because of the expense involved will greatly restrain the number of tests that can be run. Two approaches can be used to assess scuffing:

1. An objective measurement such as percent scuffed area, engine torque loss, engine blow-by flow rate.

2. A subjective measurement such as severity rating (1 = very bad to 10 = very good, for instance), rank order of severity of scuffing and use rank number as data.

Factor selection and levels.   This case was discussed at great length to review the possibilities for reducing the actual number of engine tests. The engine was a four-cylinder engine with sleeves, which allowed each cylinder to become a test site in itself, thus accomplishing four trials with one engine run. However, because of possible cylinder-to-cylinder variation, the experiment should be blocked on cylinders (treat the cylinder location as a four-level factor, factor K). To reduce the overall experimental time it was suggested to use two engines for the trials, one engine could be on the test stand while the other was being prepared for the next trial. Blocking again was recommended for the engines and treat engines as a two-level factor, factor L. The factors that were thought to influence scuffing were: A, piston/sleeve fit; B, piston ovality; C, piston contour; D, piston plating thickness; E, piston finish; F, piston material; G, piston cooling slot; H, sleeve finish; I, ring design; and J, engine rating. The engine rating, base and uprate, is a function of the head design and fuel control system, which are interchangeable on the basic block.

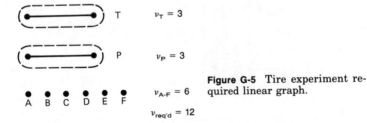

$\nu_T = 3$

$\nu_P = 3$

$\nu_{A-F} = 6$

$\nu_{req'd} = 12$

Figure G-5 Tire experiment required linear graph.

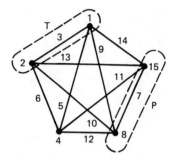

Figure G-6 L16 standard linear graph a.

Assignment of factors. All of the factors plus the blocking items require 14 degrees of freedom, so an L16 OA is required. To make it easy to visualize the trial conditions, the engine rating was assigned to column 1 and columns 2, 4, and 6 were merged to form a four-level column for cylinder number. The remaining factors were assigned to the columns as shown in Table G-6.

Example G.6 Manual Transmission Lube System

A manual automotive transmission typically is lubricated by the rotating gears splashing oil throughout the entire mechanism. The oil level in the transmission is high enough to submerge the lower portion of some of the gears to accomplish splash lubrication. Other devices in the transmission may direct the

TABLE G-5 Tire Experiment Factor Assignment

					Factors							
	T	*A*	*B*	*C*	*P*	*D*	*E*	*F*				
					Column no.							
Trial no.	1-2-3	4	5	6	7-8-15	9	10	11	12	13	14
1	1	1	1	1	1	1	1	1	1	1	1
2	1	1	1	1	2	2	2	2	2	2	2
3	1	2	2	2	3	1	1	1	2	2	2
4	1	2	2	2	4	2	2	2	1	1	1
5	2	1	1	2	3	1	2	2	1	1	2
6	2	1	1	2	4	2	1	1	2	2	1
7	2	2	2	1	1	1	2	2	2	2	1
8	2	2	2	1	2	2	1	1	1	1	2
9	3	1	2	1	3	2	1	2	1	2	1
10	3	1	2	1	4	1	2	1	2	1	2
11	3	2	1	2	1	2	1	2	2	1	2
12	3	2	1	2	2	1	2	1	1	2	1
13	4	1	2	2	1	2	2	1	1	2	2
14	4	1	2	2	2	1	1	2	2	1	1
15	4	2	1	1	3	2	2	1	2	1	1
16	4	2	1	1	4	1	1	2	1	2	2

TABLE G-6   Piston Scuffing Experiment Factor Assignment

													Factors
	J	K	A	B	C	D	E	F	G	H	I	L	
												Column no.	
Trial no.	1	2-4-6	3	5	7	8	9	10	11	12	13	14	15
1	1	1	1	1	1	1	1	1	1	1	1	1	1
2	1	1	1	1	1	2	2	2	2	2	2	2	2
3	1	2	1	2	2	1	1	1	1	2	2	2	2
4	1	2	1	2	2	2	2	2	2	1	1	1	1
5	1	3	2	1	2	1	1	2	2	1	1	2	2
6	1	3	2	1	2	2	2	1	1	2	2	1	1
7	1	4	2	2	1	1	1	2	2	2	2	1	1
8	1	4	2	2	1	2	2	1	1	1	1	2	2
9	2	1	2	2	2	1	2	1	2	1	2	1	2
10	2	1	2	2	2	2	1	2	1	2	1	2	1
11	2	2	2	1	1	1	2	1	2	2	1	2	1
12	2	2	2	1	1	2	1	2	1	1	2	1	2
13	2	3	1	2	1	1	2	2	1	1	2	2	1
14	2	3	1	2	1	2	1	1	2	2	1	1	2
15	2	4	1	1	2	1	2	2	1	2	1	1	2
16	2	4	1	1	2	2	1	1	2	1	2	2	1

oil to appropriate locations to provide uniform lubrication and cooling of components.

Problem.   To evenly distribute the lubrication oil on critical components of a manual transmission.

Objective.   To achieve an average lubrication score of 25.0 or greater and to equally distribute the lubrication (no individual location below a value of 3.0).

TABLE G-7   Factors for Manual Transmission Lube Experiment

Factors and levels

A_1 2nd range at 3000 rpm
A_2 3rd range at 3000 rpm
A_3 4th range at 2000 rpm
A_4 5th range at 2000 rpm

B_1 Sealed bearing
B_2 Standard bearing

C_1 No splash assist
C_2 Splash assist

D_1 New trough design
D_2 Old trough design

Measurement system.   The transmission is operated in the various ranges at specific speeds and the amount of lubrication flowing to a location observed through plastic observation ports. The amount of lubrication flowing to a location is rated on a 0 to 5 scale (0 = none, 3.0 adequate, and 5 = abundant). The seven locations requiring lubrication are then totaled for an overall transmission score.

Factor selection and levels.   This experiment is set up with some noise factors included in the inner array, although this is not recommended. Four different gear ranges were selected with specific input speeds used in those ranges. A bearing location could use either a sealed or standard flow through design and a trough for directing lubrication could be of an old or new design. The factors and levels are shown in Table G-7.

Assignment of factors.   A desire to investigate the $A \times B$, $A \times C$, and $A \times D$ interactions adds some interesting considerations to the column assignments. Figure G-7 shows the linear graphs required to evaluate the interactions when one of the factors is four levels. The modified L16 OA is provided in Table G-8 along with the assignment of factors and interactions.

Conducting the experiment.   The transmission specified by the trial condition is placed in the particular range for the test and the amount of lubrication observed at the seven critical sites. The total of the lubrication scores for the seven sites is also indicated in Table G-8. The individual scores were also

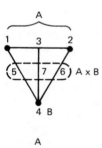

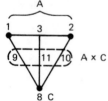

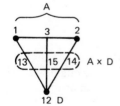

Figure G-7 Lubrication experiment linear graphs.

TABLE G-8 Lube Experiment Factor Assignment

					Factors			Lube score
	A	B	A×B	C	A×C	D	A×D	
					Column no.			
Trial no.	1–2–3	4	5–6–7	8	9–10–11	12	13–14–15	
1	1	1	1 1 1	1	1 1 1	1	1 1 1	27.5
2	1	1	1 1 1	2	2 2 2	2	2 2 2	31.5
3	1	2	2 2 2	1	1 1 1	2	2 2 2	27.0
4	1	2	2 2 2	2	2 2 2	1	1 1 1	27.5
5	2	1	1 2 2	1	1 2 2	1	1 2 2	22.0
6	2	1	1 2 2	2	2 1 1	2	2 1 1	30.5
7	2	2	2 1 1	1	1 2 2	2	2 1 1	25.0
8	2	2	2 1 1	2	2 1 1	1	1 2 2	28.5
9	3	1	2 1 2	1	2 1 2	1	2 1 2	26.5
10	3	1	2 1 2	2	1 2 1	2	1 2 1	29.5
11	3	2	1 2 1	1	2 1 2	2	1 2 1	21.5
12	3	2	1 2 1	2	1 2 1	1	2 1 2	29.0
13	4	1	2 2 1	1	2 2 1	1	2 2 1	27.0
14	4	1	2 2 1	2	1 1 2	2	1 1 2	29.5
15	4	2	1 1 2	1	2 2 1	2	1 1 2	21.0
16	4	2	1 1 2	2	1 1 2	1	2 2 1	25.0

recorded and later analyzed, but for this example only the total score is addressed.

Analysis and interpretation. The analysis of variance for the lubrication score is summarized in Table G-9, and a pooled version of the ANOVA is summarized in Table G-10. The items that were pooled in the original table were all in the 6.00 or less range, while the two items that stand out the largest were the use of the unsealed bearing design and the gear splash assist. These two items contributed over half of the variation in the observed results. The estimate for the average of the $B_1 C_2$ condition is:

$$\hat{\mu}_{B_1 C_2} = \overline{B}_1 + \overline{C}_2 - \overline{T}$$

$$= 28.00 + 28.88 - 26.78 = 30.10$$

The confidence interval for this condition is

$$CI = \sqrt{(F_{.10;1;13} \times V_e)/n_{eff}}$$

$$= \sqrt{(3.14 \times 4.35)/(16/3)} = 1.60$$

The average lubrication score is then estimated to fall somewhere in the range of 28.5 to 31.70 with 90 percent confidence. The objective of the experiment was to achieve an average of at least 25.0, so that objective potentially can be met with those two design features.

TABLE G-9 ANOVA Summary for Lubrication Score

Source	SS	ν	V
A	15.92	3	5.31
B	23.77	1	23.77
C	70.14	1	70.14
D	0.39	1	0.39
$A \times B$	16.68	3	5.56
$A \times C$	9.06	3	3.02
$A \times D$	14.55	3	4.85
Total	150.48	15	

Confirmation experiment. The confirmation experiment is necessary to verify the conclusions of the screening experiment. Several transmissions should be assembled with the unsealed bearing and splash assist and tested to prove the theory.

Example G.7 Spot Welding Development

During the process of spot welding, pieces of sheet metal are joined together by forcing a high current through a localized area of contact; the thin metal is often distorted for various reasons. Often parts are held in jigs for locating purposes, but the heating and pressures used to hold parts still result in a warped workpiece. A research and development proposal of performing the spot welding through a sealant material was thought to make the distortion potentially worse.

Problem. Distortion of part during the spot welding process.

Objective. To minimize the distortion of experimental sheet metal components during spot welding to a value of less than .75 mm on the average and 1.00 mm for any individual value.

Measurement system. A height gauge is used to measure the right and left ends of components held in a fixture prior to and after welding to obtain welding distortion. The welding fixture is shown in Fig. G-8.

TABLE G-10 Pooled ANOVA Summary for Lubrication Score

Source	SS	ν	V	F	P
B	23.77	1	23.77	5.46‡	12.90
C	70.14	1	70.14	16.12$^{\#}$	43.72
e_p	56.58	13	4.35		43.38
Total	150.48	15			100.00

†At least 90% confidence.
‡At least 95% confidence.
$^{\#}$At least 99% confidence.

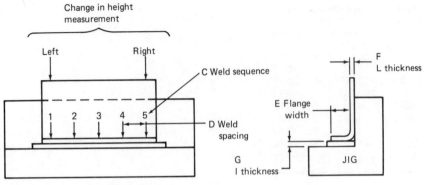

Figure G-8 Spot welding experiment.

Factor selection and levels. The factors and levels selected for the welding experiment are shown in Table G-11. The weld schedule includes two different arrangements of welding contacts, welding pressure, and amperage.

Assignment of factors. Since there are 7 two-level factors, the experiment can use an L8 OA with all columns assigned.

Conducting the experiment. The experiment was accomplished in two blocks, since changing the welding schedule was somewhat difficult. The remaining items are very simple to change from trial to trial and were completely randomized within the blocks. Five parts were manufactured under the trial conditions, and the change in height measured at both the right and left ends before and after welding. The resultant information is summarized in Table G-12, and the data is coded by multiplying each of the deflection values by 100 to allow computer analysis.

Analysis and interpretation. A series of ANOVA summary information is shown in Tables G-13 to G-17, which include the unpooled and pooled versions of the ANOVA of raw data and S/N data, in that order. The analysis of the raw data indicates that up to five of the seven factors contribute something to the variation in deflection; however, two of the factors make a very small contribution as

TABLE G-11 Spot Welding Factors and Levels

		Levels	
Factors		1	2
A	Weld schedule	1	2
B	Sealant thickness	3 mm	5 mm
C	Weld sequence	1-2-3-4-5	1-5-2-4-3
D	Weld spacing	20 mm	30 mm
E	Flange width	10 mm	15 mm
F	L thickness	.75 mm	1.00 mm
G	I thickness	1.00 mm	1.5 mm
N	Measurement position (noise)	Left	Right

TABLE G-12　Spot Welding Experimental Data

Trial no.	Left end					Right end					S/N
1	100*	97	76	87	80	82	71	59	68	58	−37.95
2	61	66	62	60	73	51	56	56	52	56	−35.51
3	79	72	74	65	67	81	82	78	78	74	−37.52
4	96	102	116	111	108	103	106	109	124	99	−40.64
5	75	89	73	79	67	75	88	68	71	67	−37.57
6	95	102	89	88	84	97	100	85	94	80	−39.24
7	109	115	107	109	110	113	111	94	106	106	−40.68
8	98	125	90	109	108	74	94	70	103	110	−39.95

*Deflection in mm × 100.

shown in Table G-14. Further pooling of the column effects as shown in Table G-15 indicates that only three factors are really contributing the majority of the variation in deflection. The first weld schedule, the thinner sealant thickness, and the thicker I component reduce the average deflection during the welding process since deflection is a lower-is-better characteristic. The ANOVA tables indicate which columns are important, but the level averages indicate which level is superior. The estimate of the mean is

$$\hat{\mu}_{A_1 B_1 G_2} = \overline{A}_1 + \overline{B}_1 + \overline{G}_2 - 2\overline{T}$$

$$= 79.88 + 75.93 + 76.90 - 2(86.53) = 59.65$$

TABLE G-13　ANOVA Summary for Spot Welding Experiment

Source	SS	ν	V	F	P
A	3537.81	1	3537.81	45.66	12.18
B	8988.81	1	8988.81	116.02	31.37
C	42.06	1	42.06	0.54	
D	510.06	1	510.06	6.58	1.52
E	72.19	1	72.19	0.93	
F	768.81	1	768.81	9.92	2.43
G	7411.25	1	7411.25	95.66	25.81
T_1	21,330.99	7			
N	627.19	1	627.19	8.10	1.93
$A \times N$	0.44	1	0.44	0.01	
$B \times N$	162.44	1	162.44	2.10	0.30
$C \times N$	980.00	1	980.00	12.65	3.18
$D \times N$	7.19	1	7.19	0.09	
$E \times N$	61.25	1	61.25	0.79	
$F \times N$	281.25	1	281.25	3.63	0.72
$G \times N$	0.81	1	0.81	0.01	
T_2	23,451.56	15			
e	4958.38	64	77.47		
T_3	28,409.94	79			

TABLE G-14 Partially Pooled ANOVA Summary for Spot Welding Experiment

Source	SS	v	V	F	P
A	3537.81	1	3537.81	45.60#	12.18
B	8988.81	1	8988.81	115.86#	31.37
D	510.06	1	510.06	6.57‡	1.52
F	768.81	1	768.81	9.91#	2.43
G	7411.25	1	7411.25	95.53#	25.81
N	627.19	1	627.19	8.03#	1.93
C×N	980.00	1	980.00	12.63#	3.18
e_p	5586.01	72	77.58		21.58
T	28,409.94	79			100.00

†At least 90% confidence.
‡At least 95% confidence.
#At least 99% confidence.

TABLE G-15 Completely Pooled ANOVA Summary for Spot Welding Experiment

Source	SS	v	V	F	P
A	3537.81	1	3537.81	31.74#	12.06
B	8988.81	1	8988.81	80.64#	31.25
G	7411.25	1	7411.25	66.48#	25.69
e_p	8472.07	76	111.47		31.00
T	28,409.94	79			100.00

†At least 90% confidence.
‡At least 95% confidence.
#At least 99% confidence.

TABLE G-16 S/N ANOVA for Spot Welding Experiment

Source	SS	v	V
A	4.22	1	4.22
B	9.07	1	9.07
C	0.10	1	0.10
D	0.33	1	0.33
E	0.01	1	0.01
F	1.24	1	1.24
G	7.93	1	7.93
T	22.90	7	

The confidence interval for this condition is

$$\text{CI} = \sqrt{(F_{.10;1;76} \times V_e)/n_{\text{eff}}}$$

$$= \sqrt{(2.78 \times 111.47)/(80/4)} = 3.94$$

The average with 90% confidence falls within the range of .5571 to .6359 mm when the data is uncoded by dividing by 100. The variance of 111.47 provides a

TABLE G-17   Pooled *S/N* ANOVA for Spot
Welding Experiment

Source	SS	ν	V	F	P
A	4.22	1	4.22	10.11‡	16.60
B	9.07	1	9.07	21.73#	37.78
C	7.93	1	7.93	19.00‡	32.81
e_p	1.68	4	0.42		12.81
T	22.90	7			100.00

†At least 90% confidence.
‡At least 95% confidence.
#At least 99% confidence.

standard deviation of 10.56 mm (.1056 mm when uncoded). If the average is as high as .6359 and three standard deviations are added to that value, the predicted value is .9527 mm, which is within the objective of the experiment. The average value and individual values are predicted to meet the experimental objectives.

An interpretation of the *S/N* data indicates that the same factors and levels increase the *S/N* value as shown in Tables G-16 and G-17. The average *S/N* value predicted for $A_1B_1G_2$ is −35.85 db. This is a situation typically encountered with either the *S/N* ratio for LB or HB characteristics, the factors which decrease or increase averages, respectively, will increase the appropriate *S/N* ratio. The analysis of *S/N* data, therefore, is not as meaningful as in the case of NB characteristics.

For the condition $A_1B_1G_2$, one can see that trial 2 has that treatment condition and the average results for only five repetitions fall within the confidence interval for the average of both the raw data and *S/N* data. Trial 2 is like a confirmation experiment within the original experiment; the raw data varies from .51 to .73 mm deflection, which gives great promise for the final confirmation experiment that may be run with a larger number of parts.

Example G.8.   Brake Material Bonding Experiment

Brake friction material is chemically bonded to a steel brake shoe to provide support for the friction material during operation. The integrity of the bonding process is critical to the function of the brake shoe assembly. The process fundamentally consists of cleaning, applying adhesive, and drying of the individual parts, which are then joined in an assembly and heated to activate the chemical bonding. At the end of the bonding process, samples of the batch of assemblies are tested for shear strength of the bond in a special test fixture. An experiment was proposed to increase the bond strength as much as possible (HB).

Problem.   Insufficient bond strength on the final assembly bond tester.

Objective.   To achieve a minimum bond shear strength of 150 lb.

Measurement system.   A section of the brake shoe assembly is cut off, mounted in the test fixture, and load applied until shear failure is induced.

TABLE G-18   Brake Shoe Bonding Factors and Levels

	Levels	
Factors	1	2
Shoe		
K Shot blast	1 min	3.5 min
A Solvent wash	Butyl alcohol	MEK
B Drying time	15 min	30 min
C Shoe adhesive	Old	New
D Drying time	2 h	4 h
Lining		
E Solvent wash	Butyl alcohol	MEK
F Drying time	30 min	60 min
G Lining sealer	Old	New
H Drying time	2 h	4 h
I Lining adhesive	Old	New
J Drying time	2 h	4 h
Brake shoe assembly		
L Clamp load	500 lb	600 lb
M Curing time	30 min	120 min
O Curing temperature	360°F	400°F
N Cooling time	30 min	60 min

Factor selection and levels. The process consists of many steps, subsequently, several factors are possible for evaluation. The 15 factors and levels chosen for the experiment are shown in Table G-18. In this situation, some of the levels of the factors cause the chosen factor to actually become a noise factor. For instance, the age of the adhesives is really a noise to the bonding process. If the age turns out to be statistically significant and contributes a large percentage to variation, then the age of the adhesives will have controlled more closely.

Assignment of factors. The factors are going to completely saturate an L16 OA, but in this case the assignment was not in an alphabetical order. The factors were assigned sequentially to columns 1 to 15 in this order: $D, J, O, B, C, A, F, G, E, I, K, L, M, N,$ and H. Since these were all two-level factors, the OA will not be shown.

Conducting the experiment. The brake shoe assemblies were made according to the trial "recipes" and shear-tested in the appropriate fixture. The load results for three different shoes in a top and bottom position are shown in Table G-19. In this situation, position is treated as a noise factor, P.

Analysis and interpretation. Two points are very clear from a cursory analysis of the data. One, there is a lot of variation within a trial; two, half of the trials registered zero or very nearly zero loads, which means one factor is substantially affecting the results. Concerning the second point, one of the levels of a factor could be outside the operating window for that factor. For this reason, care must be taken not to select levels too widely apart. A pooled ANOVA summary is shown in Table G-20. The one factor that controls nearly half of the observed variation is M, the final curing time, and the second level provides substantially superior loads relative to the first level. The 30-minute cure time is obviously

TABLE G-19 **Brake Shoe Assembly Shear Force Results**

Trial no.	Top			Bottom		
1	0*	0	0	0	0	0
2	40	74	52	12	57	15
3	23	8	14	61	9	70
4	0	0	0	0	0	0
5	0	0	0	0	0	0
6	106	122	68	92	131	47
7	9	96	64	11	32	11
8	0	8	9	0	3	3
9	124	165	69	82	52	11
10	0	6	6	0	3	4
11	0	6	4	0	6	5
12	81	73	51	67	8	21
13	9	123	21	103	6	22
14	0	0	0	0	0	0
15	0	1	0	0	0	0
16	7	11	26	62	68	65

*Shear force, lbs.

inadequate, but some other cure time between 30 and 120 minutes may increase shear strength suddenly. The average shear strength under the 120-minute cure condition was only 53.27 lb, which means the experimental objective cannot be met with only one factor unless an increased cure time beyond 120 minutes causes a dramatic rise in shear strength.

Another note on the ANOVA results, the error variance of the experiment is quite large. The standard deviation is 23.75 lb, which provides a six-sigma span of 142.5 lb for the measurement discrimination. A recommended follow-up from this experiment would be to investigate the repeatability of the shear test machine to reduce the experimental error as much as possible. The variability of

TABLE G-20 **Pooled ANOVA Summary for Brake Shoe Experiment**

Source	SS	ν	V	F	P
O	5148.02	1	5148.02	9.13#	3.22
B	4043.01	1	4043.01	7.17#	2.44
F	1708.59	1	1708.59	3.03†	0.80
E	4746.09	1	4746.09	8.41#	2.94
I	2137.59	1	2137.59	3.79†	1.11
M	64,740.09	1	64,740.09	114.78#	45.10
N	3492.09	1	3492.09	6.19‡	2.06
O×P	4069.00	1	4069.00	7.21#	2.46
N×P	3712.59	1	3712.59	6.58‡	2.21
e_p	48,507.18	86	564.04		37.66
T	142,304.25	95			100.00

†At least 90% confidence.
‡At least 95% confidence.
#At least 99% confidence.

the measuring device could be masking the presence of some other contributing factors as indicated by the low percentage contributions; recall that error variance effectively subtracts from the factor sum of squares. If the measuring device has good repeatability, a high S/N ratio, then there are other factors not included in the original screening experiment that are contributing to the observed variation. In addition, the repeatability of the shear tester may not be the only problem with it. The calibration should also be checked, since that may account for the low loads obtained.

Summary.   This example was included to demonstrate the kinds of problems encountered in experimentation. Not all experiments succeed, but the two most common reasons are the lack of repeatability of the measuring system(s) used in obtaining the results and the exclusion of influential factors due to being unaware or choosing to minimize the size of the experiment by reducing the number of factors.

Index

Accumulation analysis, 154,
  161–164
Additivity, 118–120
Alpha mistake, 44–45, 60, 93–94
ANOVA, 23–24, 43, 45, 47, 128,
  204
  for a dummy treatment, 110–112
  for four-level factors, 103
  no-way, 24–31
  one-way, 31–47
    excluding the mean, 37–47
    including the mean, 32–37
    orthogonal array: L4, 76–78
    L8, 90–92
  three-way, 54–60
  two-way, 47–53
Assignment of factors and interactions, 75, 78–85, 133–135,
  139–140, 142–143, 204
Attribute data, 151–153
  analysis summary, 153–154
  multiple-class analysis, 161–164
  sample size for, 153
  two-class analysis, 157–159
Attribute data transformation, 161

Beta mistake, 60, 93–94
Blocking, 86–88

Carbonitriding, 179–180
Casting hardness, 47
Cause-effect diagrams, 71–73, 96–97
Central limit theorem, 40–42
Class weight, accumulation analysis,
  162–163
Classes analyzed, number of,
  163–164
Clutch plates, 21
Column effects method, 128, 204

Columns, merging, 102, 109, 145
Combined factors, 137–142, 145
Component search, 146–149
Confidence, meaning of, 44–46
Confidence intervals, 115, 120–124
  confirmation experiment, 123–124
  existing condition, 121–122
  predicted condition, 123
Confirmation experiment, 94–95
Confounding, 80–84
Control factor, 168–171, 176,
  185–186, 192–197, 200–201, 204
Cost constant, k, 9
Cracks, casting, 154–156, 160–162
Crosby, Philip, 2

Degrees of freedom, 28–29, 204
  for error, 29
  factor, 36–37, 40, 52
  interaction, 53
  mean, 29, 39
  total: excluding mean, 39
    including mean, 36
Die cast component, 184–191
Dummy treatment, 109–112, 145

Electronic circuit, 177–179
Experimentation:
  secondary rounds of, 129–131
  steps in, 203–205

F test, 43–47
  limitations in method, 60
Fishing reel, 180–183
Flashlight, 3
Flowcharts, 71–73, 96–97, 204
Frequency distribution, 7–8, 13, 20

Goalpost philosophy, 2–4

Hadamard, Jacques, 70
Heat treatment process,
    carbonitriding, 179–180
Hinge, automobile hood, 193–195
Histogram, 8–9
Hood, automobile, 3–5

Idle column method, 142–145
Inner noise, 169, 177, 195, 203
Interaction, 48
    linear graphs, 78–80
    multiple-level factor, 109
    plot, 54, 182, 189
    triangular tables, 79–80
Interaction columns, 75–78

k, cost constant, 9

Level(s), 31, 64, 204
    multiple, 101–103
Linear graphs, modification of,
    84–85
Loss function, 3–20
    higher-is-better characteristic,
        18–19
    lower-is-better characteristic,
        17–19
    nominal-is-best situation, 5–6,
        9–12, 16–19

Mazda, 16
Mean, estimating the, 118–120
Measurement methods, 96–97, 179,
    195–201, 203

Nested experiments, 133–137
Noise factor, 168–171, 173, 175–176,
    185–186, 188, 191, 196–197,
    200, 204
Normal distribution, 7–8

Objective, experimental, 96, 203
Observation, 25
Observation method, 126, 204
Off-line QC, 18, 169, 203
Omega transformation, 124–125
On-line QC, 16, 18, 169, 203
Orthogonal array, 63, 70, 74, 76–77,
    204
    inner, 171, 184, 191–194, 197
    outer, 171, 175, 183, 191–194,
        197
    three-level, 74

Orthogonal array, (Cont.):
    two-level, 74, 76–77
Outer noise, 169, 177, 195, 203

Parameter design, 167–172,
    175–179, 182–183, 188,
    191–197, 202–203, 205
    case studies, 176–195
    measurement system, 195–201
    strategy, 175–176
Percent contribution, 115–117
Plackett and Burman (British
    statisticians), 70, 80
Polyethylene film, 5–7
Polynomial decomposition, 106–109
Pooling, 93–95
Popcorn experiment, 95
Primary table, 185
Problem, statement of, 96, 203
Product noise, 169, 182

Quality, definition of, 1
Quality circle, 2
Quality countermeasures, 202–203

Randomization, 86
    complete, 86–87
        within blocks, 87
    effect on error variance, 88
    simple repetition, 87
Ranking method, 127
Repeatability, 196, 199
Repetitions, 170
Resolution, 81–84
Risk, alpha (see Alpha mistake)

Sample size, 89
    effective, 123
Secondary table, 185
Selection:
    of factors, 71–73, 97, 203–204
    of number of levels, 73–74, 97, 204
    of orthogonal array, 75, 204
Signal factor, 196–197, 200
Signal-to-noise ratio, 172
    calculation, 173–175, 189, 195
    characteristics: higher-is-better,
        173
        lower-is-better, 172
        nominal-is-best, 172, 175
Speed sensor, 199–201
Sums of squares:
    due to mean, 26–29, 33
    error, 27–28

Sums of squares, (*Cont.*):
  factor, 34–35, 49–50
  interaction, 51–52, 57–58
  total, 26–29, 33, 36, 38
    excluding the mean, 38
    including the mean, 26–29, 33, 36
System design, 168, 202

Tertiary table, 185
Test strategies, 63–71
  factorial, 67–68
  fractional factorial, 68–71
  many factors: all-together, 64–65, 67
  one-at-a-time, 64–66
  one-factor-at-a-time, 64–66
  typical, 63–67
Theft, 6

Tolerance, factory, 17
Tolerance design, 167–169, 178, 182, 201–203, 205
Tool wear, 7, 12–16
Transmission:
  automatic, 16
  noisy, 147–149
Trial, 64

Variable data, 151
Variance, error, 29–30
  column estimate, 91
Variation in an experiment, 89

Wear measurement, 196–199
Windshield washer pump, 24–29, 31–37, 39–41, 46

About the Author

Phillip J. Ross is an international consultant and instructor in statistical methods for component development. He holds a bachelor of mechanical engineering from General Motors Institute.